SOMETIMES THINGS
JUST HAPPEN

K. ARNOLD

authorHOUSE®

AuthorHouse™
1663 Liberty Drive
Bloomington, IN 47403
www.authorhouse.com
Phone: 833-262-8899

Published by AuthorHouse 07/22/2022

ISBN: 978-1-6655-6591-2 (sc)
ISBN: 978-1-6655-6590-5 (e)

Library of Congress Control Number: 2022913587

Print information available on the last page.

CONTENTS

CHAPTER 1

―― ∞ ――

Early Life

Our earth is a small to medium sized planet orbiting an undistinguished star. Our sun is one of billions in our galaxy and there are more than 50 billion galaxies in the universe. Objectively, our earth cannot be significant, a spec in an unimaginable expanse of space. Of course, we are not objective, and we will focus on life on our small planet.

This chapter begins with the Big Bang followed by the origin of living things on planet earth. It ends billions of years later with a few species of multi-celled organisms. It covers two thirds of the time that life has existed on earth. Along the way, the earth also changes.

The First Big Bang

It started with a bang almost 14 billion years ago. Before the bang there was nothing; after the bang there were the ingredients for our universe. Some scientists say that, maybe there was a little something before the bang, perhaps enough to fill a teacup. Others propose a cosmos of membranes... the Big Bang phenomenon occurred when they collided. But then where did the membranes come from? It hardly matters, at some point, something came from nothing (or next to nothing). That is an act of creation; therefore, there must be a creator or a God. In fact, the act of creation defines God for many people. That's about all we know for sure about God, religion deals with the rest.

After the bang, the expanding mass began to cool; initially small and then large amounts of hydrogen and helium were formed. These hot gasses began to agglomerate and eventually formed stars. The stars are the nuclear physicists of the universe; they fuse hydrogen, helium, and their fusion products into all the known elements. This fusion takes place at extraordinary high temperatures and releases massive amounts of heat. When stars exhaust (no more material to fuse), they sometimes violently explode (super nova) and their contents are redistributed throughout the cosmos. Other stars are less dramatic, they just go dark.

Astronomers have made tremendous advances in understanding the universe with the aid of advanced measurements including observations from the Hubble Space Telescope. However, some effects can only be explained by assuming there is much more matter and energy than we can detect. Therefore, some current theories propose that there is a large amount of 'dark matter' which does not emit or reflect light and does not interact with what we think of as ordinary matter. What we thought was real is a fraction of "what is" and "what is" cannot be directly measured or observed.

A good scientific theory does two things: accurately describe known observations with a minimum of arbitrary constants and provides a basis for predictions within its validated range. For example, gravity must have certain strength for matter to come together and form stars and planets. If it differed by just a bit, the universe would not have stars and planets; it would still be expanding as particles. Other properties such as the energy density of empty space, the strength of electromagnetic forces, and the mass ratio of the conversion of hydrogen to helium, must be the way they are. If these laws were not precisely defined with parameters in a narrow range, they would predict the universe would either not exist or be a very different place. These fundamental properties of the universe are required for life, as we know it, to exist.

Physicists are trying to develop a unified theory of things that would utilize quantum mechanics (which works well for small particles) and Einstein's general relativity theory (which explains the effect of gravity on large particles). If this were to happen, there should be fewer parameters and the combined theory would have more credibility. One proposed "theory of everything" provides continuity of the laws of physics and further suggests the big bang was the aftermath of the collisions of two parallel universes. It also eliminates the zero-time singularity, which pleases the mathematicians. However, the theory exits in 11-dimensional space, and it implies millions of universes, each having different laws of physics. If there were other universes, they have either disappeared or we haven't yet discovered them. It seems to most of us it would be hard to miss an entire universe.

To laymen, this is madness. It explains neither when, where, or how things first began. It seems the best science raises more questions than we had before. An alternative is that the properties of our universe are not random and that things were set up that way. While that's beginning to sound like religion again, it doesn't feel too bad compared to the alternatives.

> *The universe was created with a burst of energy; that energy transformed our galaxies, stars, and all other cosmic debris into matter. Man's mind cannot comprehend the beginning of time or the formation of the universe. We know of no feasible explanation, other than that it was the act of a creator or God. But what was God doing before he formed the universe and why did he bother? We need to keep looking.*

As we will later see, religion comes naturally to man, science does not. The scientific process is a discipline that must be learned; it is the best way we know to discover the truth about our natural world. It is a discovery process that involves hypothesis and observations and/or experiments designed to validate the hypothesis. It is our way to connect cause and effect and to test models of how things work. To have validity, the observations must be repeatable by other scientists, some of whom may have very different views; and who would like nothing more than to invalidate a competing hypothesis.

The work of science is frustrating. New ideas are commonplace, but most are wrong. Most flashes of insight lead nowhere. Experiments can be tedious and often produce negative results, or worse yet, ambiguous results. Most scientists never discover anything of significance;

they must satisfy themselves filling in gaps and testing their hypothesis and those of others.

Science cannot be used to categorically prove a causal relationship. It is always possible that there are other factors, unknown to the investigator that impact a particular outcome. Thus, hypotheses are not proved; they are validated under specified conditions. When validated under many different conditions, the hypothesis becomes accepted as a good theory. The "laws" of science or physics are widely applicable and should be accepted by essentially everyone. But that's not always the case...even Newton's law of gravity was found (by Einstein) to be invalid under some circumstances. Scientists believe and use scientific theories because they offer the best explanation of what they observe; and they are often useful.

While we can attribute a certain amount of virtue to the scientific process, scientists, in general, cannot claim that same virtue. They are just like other groups of educated people; they can produce extraordinary and innovative results. And, they sometimes lie, steal, and harbor false beliefs. There are many examples of scientist's group think and lemming-like behavior; Often there are bitter debates on the meaning of experimental results; and outright fraud in defense of reputation and/or personal advancement.

Most scientist spend half their careers learning their specialties and sometimes know little else. To win the scientist must make a significant discovery and then wait for many years to be recognized for his contribution. Fail to discover something useful and he is hardly known even within his field. Career pressures, the need for recognition, and narrowness of his background, tends to bias the scientist into thinking that whatever he is studying is profound and he alone knows the real story. And, run for the hills when you hear that a consensus of scientists agrees on this or that. Science is not about consensus. Most great scientific breakthroughs defied the consensus of their peers when they were first proposed.

Man needs explanations for things he cares about but does not understand. Further, the explanations often become embodied in his culture. Early on, man knew little about his natural world and his cultural explanations were rich and diverse. By definition, supernatural explanations for things that man does not understand are a part of his religion. As man's knowledge expands, more and more natural phenomena became understood through science and logic. Often new knowledge is resisted when it conflicts with cultural beliefs and the hardest beliefs to dislodge are religious beliefs. Regardless, the direction is clear: over time, scientific understanding displaces religious explanations of natural phenomena.

There should be no real conflict between science and religion. Religion should be about morality, intent, and purpose that endures over time. Scientific concepts are bound in time. Today's state of understanding was incomprehensible yesterday, and tomorrow will bring new directions. It is unfortunate that some scientific beliefs (unproven hypothesis) are held so tenaciously that you would think they were rooted in religion.

> *Man seeks out explanations for physical phenomena and will use supernatural explanations if others are not available.*

Today, we use science to learn about the universe; this understanding comes in spurts as scientists argue out the meaning of their observations. As our knowledge increases it sometimes differs with our embedded religious beliefs. Conflict is unavoidable.

Small Planet, Giant Universe

Our galaxy is undistinguished among the billions that exist in the universe. And our sun is only one of 200 billion stars in our galaxy. That should humble us. It is hard to imagine that the story of life on earth is particularly significant (except to us) unless it is a pattern that occurs elsewhere. And if the pattern occurs elsewhere, does it produce the same results? Scientists only speculate on these questions. Dreamers do just as well. While a great deal is now known about life on our tiny planet, there are many knowledge gaps. Many scientists believe the entire story is known. But that is only a belief they have, their own kind of religion.

About 5 billion years ago (bya) clouds of interstellar dust and gases were drawn together by gravity, and our solar system was born…just as stars have formed in billions of other places. When the matter congealed, and conditions were right, thermo-nuclear reactions ignited the furnace at the heart of the sun. These nuclear reactions continue. If they did not, even for a few moments, all life on earth would be extinguished.

About 4.6 bya the planets of our solar system formed

from the accretion[1] disk orbiting the sun; there were many more planets than exist today. They circled the sun in chaotic, relatively unstable, orbits. The newly formed earth collided with another proto-type planet about the size of mars. This is an unimaginable event that dissipated more energy than man can conceive, ripping off land masses the size of continents, and shattering them into billions of pieces. While massive amounts of solid material spun into space, most of it did not escape earth's gravity. Again, accretion occurred; this time our moon took shape circling the earth.

The collision was a glancing blow and the earth ended up with a tilt and a rotation. The size and position of our moon stabilizes the earth's rotation. The earth's rotation gives us days and night; the earth's tilt gives us our seasons. Both are necessary for the successful development of most living things. The moon also slowed the earth's rotation speed. Initially days were only 6 hours long; days became longer as the earth's rotation speed decreased. Days will continue to get longer and nights shorter…neither will affect out sleep habits anytime soon.

For the next 500 million years, to about 4 bya, comets, asteroids, meteorites, and other celestial debris relentlessly slammed the recently reformed earth. This bombardment was an apparent one-time event resulting from a bobble in the asteroid belt. We can get a sense for the number and size of these collision events by looking at the craters on surface of the moon. Earth would look the same except, over time,

---

[1] Accretion is the gradual accumulation of smaller chunks into larger ones. In space, most of the material in the same general gravitational orbit eventually forms into a single large mass.

work and gave a shocking new view of the surface of our planet under the oceans.

The earth's surface is not a uniform layer but many large "plates" that butt up against each other. The movement of one plate against another results in earthquakes, the collision of land masses, and the formation of mountain ranges. Without plate tectonics, the earth's surface would appear to be flat and covered with water over a mile deep. Plate movement is an ongoing renewal process as one plate rides over another, exposing new material and burying the old.

Early on, land movements were frequent and dramatic compared to today. Land masses are now moving to recombine into larger continents The last great land movement resulted in the formation of the Himalayan mountains; India collided with the Asian continent a mere 50 mya. Africa is crashing into Euro/Asia and the Atlantic Ocean is disappearing recombining the North America's with Europe. However, the current rate of movement is about one inch a year. (It's not yet time to learn French.)

> *In addition to the fundamental properties of the universe, there were many other things that had to exist at the right time and place for us to exist. A stable orbit at the right distance from the sun; a molten core; oceans of water; plate tectonics, etc. If the purpose of the universe was to produce us, then getting everything right would be truly miraculous. Perhaps it's presumptuous to assume that we are the purpose of the universe.*

Conditions for Life

On earth there is a cooperative distribution of basic elements. Keep in mind where elements came from…the grand mixer was and is the life cycle of stars. After earth formed, we had some additional shaking from collisions of space material within the solar systems. Fortunately, earth got a heavy dose of carbon, a shamelessly promiscuous element (it reacts with almost everything), necessary for the formation of organic compounds. We also have plenty of hydrogen, oxygen, nitrogen, and bits of rare materials needed for the complex organic chemistry that occurs in living organisms.

Most materials we encounter are not elements but compounds made up of several elements combined into a single molecule. For example, water is a combination of hydrogen and oxygen; clearly compounds may have very different characteristics than their components. Compounds form when the raw materials (elements or other compounds) come together under the right conditions. Chemistry is the science that deals with the combination of these elements and the conversion of one material to another. Organic chemistry deals with the chemical reactions involving that wench carbon. As we will see, life is about organic chemistry. Shortly after the earth formed there was plenty of water, carbon dioxide, methane, and ammonia, but no free oxygen. Chemists call this a reducing, or anaerobic environment.

Some organic compounds form naturally under anaerobic conditions. For example, some amino acids and nucleic acids would form under the anaerobic conditions that existed on earth 3.6 bya. Also, membranes and simple proteins would occasionally form. Many of these reactions are anaerobic; in fact, free oxygen tends to destroy complex organic molecules. Enzymes are also involved in most organic chemical reactions. They are proteins that catalyze or accelerate the reaction rates but are not used up as are raw material. Enzymes provide the kick-start and are the enablers of rapid chemical reactions.

The chemistry required for life involves several types of reactions:

- Simple materials like carbon or carbon dioxide are metabolized to release the energy needed for other reactions.
- Various complex proteins are formed from organic building blocks like amino acids and simpler proteins.
- Nucleic acids are duplicated providing a means for replicating organic materials.

Essentially all of these reactions require one enzyme or another. Catalyzed organic chemistry just happens… consistent with the laws of chemistry. It is not an indicator of life. However, there would be no life without it.

DNA (deoxyribonucleic acid) is a nucleic acid with two complimentary strands that contains genetic code. It replicates by first separating into the two strands with each serving as a template for the other. Complimentary organic bases are naturally attracted to the DNA strands and are snapped into place along both strands. What usually results are two identical DNA molecules where before there was one. DNA replication is the key to the stable inheritance of traits and the means of passing them from one generation to the next. However, errors occasionally occur in the replication process. For example, if an amino acid is inserted into the chain in the wrong place, a DNA molecule will form with different properties. These "mutant" DNA segments occur on a random basis; and are essential for the adaptation of life to an ever-changing environment.

Water is essential for all life. It is truly a remarkable material: the media for all marine life; a reactant in the fundamental chemistry of life, and the major component of all multi-celled (plants and animals) organisms. Likewise, life could not exist without an abundance of carbon. A third material, oxygen, plays an equally important role, but in a less obvious way.

Oxygen, when dissolved in water, becomes a powerful "oxidizer" which react with many materials. For example, iron reacts to form iron oxide, commonly known as rust. Carbonaceous rocks react to form limestone. Dead organic matter rots to form carbon dioxide, methane, and other compounds. When a material burns, it is oxidized, combining oxygen with the material being consumed to form mostly carbon dioxide or carbon monoxide. Dissolved oxygen attacks essentially all organic material, usually breaking down complex proteins into smaller molecules. This is a profoundly serious matter for living organisms, all of which are based on (organic) proteins. Unless they are protected, essential all proteins will decompose.

Why then is oxygen such a key material when it has such a destructive potential? The reason is that metabolism based on oxygen is very efficient relative to other metabolic reactions. Without oxygen-based metabolism, it is unlikely that large multi-celled animals could exist. Oxygen levels in the earth's atmosphere have varied from essentially zero, stepping up to as high as 35% and settled back down to current levels of about 20%. As we will see these changes had an enormous effect on how and when various life forms existed and prospered.

> *Life could not form without carbon, water, and oxygen. How many places in our universe are these three materials likely to coexist under conditions similar to those on earth? Billions and billions.*

What is Life?

Why do we study life? It may be a rather insignificant aspect of our huge universe. If it were to disappear, our earth and solar system would remain pretty much the same and certainly the universe would be totally unperturbed. We study life because we are a part of it. Are our studies more than just self-interest?

They certainly are! Life is the most complex thing we know in the entire universe. The proteins that make up living organisms are orders of magnitude more complex than inorganic materials. An organism existing in complex social structures is off the complexity scale of anything we know of anywhere else in the universe. An amoeba is more complex than a star...and a lot less predictable. An astronomer can predict what a star will do a million years from now; a biologist can only guess what an amoeba will be doing a minute from now.

Observing living organisms is the most interesting thing we know to do. There are millions of species with thousands of new ones reported each year. Each one has its own special place in the overall scheme of things. And life exists everywhere on earth, in every conceivable niche. Each life form has its own special traits that allow it to survive in its environment. And those traits can change as environmental conditions change.

Life would be fascinating even if it were strictly utilitarian. But it is more. It provides the color, shape, texture, and movement that transform our environment into an awesome place. If we somehow were not a part of life and it went away, we would have nothing of interest or beauty. Do ants think life is fascinating? Probably not. It takes organisms with emotions and cognition to experience the magic of life.

So, what is life? If we used conventional definitions for life, we would miss its early formations. For much of our recent history, men thought there was a "vital force" that distinguished the living from the non-living. As we learn more and more of the minute details of the structure and composition of living organisms, it becomes apparent that living organisms are made up of the same stuff as nonliving materials; the materials are simply organized differently. Therefore, a definition for life based on chemical composition is not possible.

Using another approach, living objects can be distinguished from others if they perform basic functions such as growth, adaptability, and reproduction. A definition based on functions works fine for most plants and animals. However, it is not useful when applied to sponges, fungi, or bacteria, which were some of the earliest life forms. As we seek the origin of life, the only useful approach is to look for the origin of inheritable traits. That is, look for organisms that encode the information needed to replicate and then pass the information to their heirs. Take for an example bacterium; passing on traits means passing on the information needed to make the specific proteins that bacteria need to live and reproduce.

When we search for the origin of life, we are seeking the universal common ancestor for all existing life forms, from the smallest bacteria to man. We believe that all current life emerged from a single ancestor because all life forms have cells with the same properties. First, the cell membranes of all organisms are lipid bilayers that effectively transport materials in and out of the cells. They all use DNA with the same four amino acids to code information and utilize the same process steps to make new proteins. Finally, all living cells utilize ATP[2] to store and transfer energy. These common traits strongly suggest all organisms have a common ancestor. Is there something special about these traits? Yes! They work. Could there be other "life" that does not have these traits. Certainly, but if it happened here, we have not yet found it.

> *The mechanics of life are driven by very complex organic chemistry; the simplest living organism is more complex than any inorganic material in the universe. We say*

---

[2] ATP is a molecule formed by cell metabolism. It can be stored and transported within the cell. It provides the energy needed to manufacture proteins.

that an organism is alive if it inherits traits from its parents. With inheritance (and adaptation), it is the most wondrous thing we know. Without inheritance, it's not life; it's just organic chemistry.

The Magic Replicator

There is clear evidence that bacteria existed by 3.5 bya. Modern bacteria are the simplest life form on earth and yet they are tremendously sophisticated molecular machines, practicing some very complex biochemistry.

They all have four basic capabilities:

- The membrane that forms the cell regulates molecular traffic into and out of the cell.
- A cell replication process that passes traits from parent to child.
- A means to synthesize proteins that catalyze chemical reactions within the cell.
- A means to convert energy into a form (ATP) useful for organic chemical reactions.

Modern day bacteria are all far too complex to have just randomly occurred. And different types of bacteria do very different things. In general terms, we understand how the necessary components might form under the conditions that existed at the time. However, just how the proteins, nucleic acids, and membranes came to interact to form a living bacteria cell remains a mystery. About 4 bya it did happen and a beachhead for life was established. Call it a "magic replicator"; it formed and, when its duplicates survived, it became the ancestor of all living organisms.

Once established, biological expansion requires that the information stored by the structure of the nucleic acid be used to form exact duplicates. The replicates that form, call them proto cells, have the same genetic coding or DNA as their parents. In turn they also replicate.

Replicators occasionally make errors (mutations) and some of mutations are also replicators…with different traits. The replicators (original plus mutants) compete for the available raw materials. Winners increase their presents relative to losers. And the process repeats. Over a long time, some mutants became very complex; however, in the eyes of the cosmos, the replicators, and all the mutants are appallingly expendable.

> *Everything we know says that all life on earth originated from a single biological event over 3.5 bya. This remarkable event occurred when a packet of stuff cleaved itself and produced an heir with the same genetic material. This is sometimes called the "Big Birth." When repeated again and again life becomes diversified.*

First Life and a Set of Rules

The magic replicator set out a fundamental set of rules that apply to all life forms. Summarizing these rules, living organisms produce "offspring" with specific traits based entirely on the information encoded in their DNA.

Occasionally, random errors occur in the replication process resulting in what are called mutations. A mutation will have slightly different DNA and somewhat different traits. While most mutants do not survive, some successfully compete and reproduce. More generally, the organisms with the best survival traits, encoded in their genetic material, preferentially survive. Thus, over time, as the environment changes, some organisms adapt and continue to prosper. Other less competitive organisms suffer and likely go extinct.

The case for the first replicator forming randomly is reasonably strong. The required materials all form naturally, and the energy needed was present in many forms. The caldron of raw materials had hundreds of millions of years to bubble and simmer for life to begin. Science will probably provide an explanation in time; but that is a projection not a given. As more is known, the formation of life becomes more comprehensible. At the same time life becomes even more astonishing!

We call this decent with modification; fitness for survival is the determining factor in what traits evolve. This could have been observed at any time over the last 3.6 billion years; it was first stated by Darwin a mere 150 years ago. It is still not accepted by some.

Once formed, living organisms can be tough; some were able to adapt and survive the most extreme insults nature has to offer. However, about 500,000 years after first life, the conditions on earth radically change in a way that destroyed essentially all first-generation life forms. Then as now, without adaptation all life would have ended.

Some magic occurs and life emerges. Replicators, by making copies of themselves, transmit information. The message is "here is a design for an organism that thrives in this environment. This is truly novel. Living organisms reverse the slow dispersion of mass and energy throughout the Universe. For this reason alone, life is special...but it's going to get a lot better.

Green is Better

Above we review how early life formed by decent with adaptation. About 3 bya, a major shift occurred and suddenly there was a better mousetrap for biological expansion. Some very different chemistry produced bacteria with some radically different traits. Cyanobacteria, sometimes called blue-green algae, is bacteria, not algae, and is not necessarily blue or green. What it invented was **photosynthesis.** It probably emerged over hundreds of millions of years; it had a transforming impact on earth. It enabled a much more rapid replication process than the one described above.

In very general terms, photosynthesis is a process that uses energy (sunlight) to produce food. Cyanobacteria like other bacteria use hydrogen in their replication process. Unlike other bacteria, cyanobacteria learned to sup off the hydrogen from a water molecule and release free oxygen as waste. This may be the single most important metabolic innovation in the history of life. It was invented by the world's first and best chemists, bacteria.

Imagine bacteria puffing bits of oxygen into an alien

world of methane, carbon dioxide, and nitrogen. Oxygen is very reactive and quickly combines with iron, the most abundant material on earth. At the time the oceans were a dirty brown color. As the iron oxidized, it coagulated and fell to the ocean floor, leaving us the beautiful blue seas. Oxygen also reacted with other minerals giving us limestone, granite, and other rock forms.

And the puffing of oxygen continued. Gradually, over 500 million years or so, the concentration of oxygen in the atmosphere rose; earth became the only planet in our solar system with an oxygen-rich atmosphere. The puffers had managed to produce enough oxygen to raise the concentration to about 1-2 percent of what it is today. That's a lot of oxygen considering that a large amount was used to rust iron.

Oxygen readily attacks proteins and was toxic to most of the bacteria that existed at the time. Adaptations occurred; some bacteria developed built-in antioxidants that could block or scavenge the oxidizing agents. Other innovations included bacteria with metabolic chemistry that consumed oxygen. A biological war was in place, with oxygen puffers producing a poison for some bacteria and nourishment for others.

The metabolic chemistry that utilizes oxygen directly is extremely efficient and, over time, organisms that used oxygen-based metabolism were clear winners. Thus, the transformation: first generation bacteria which evolved under anaerobic conditions were replaced by more robust bacteria which thrive on oxygen. The original anaerobic bacteria took cover or went extinct; cyanobacteria thrived and still exist today.

> *Over half of the time that there has been life on earth, only bacteria existed. They did a big job by surviving very hostile conditions, developing complex forms, and then converting the initial poisonous environment to one with free oxygen. Bacteria transformed earth from Dante's Hell to a place somewhat friendly for life like you and I. And for this they get little respect.*

Age of The Microbes

(A microbe is a microscopic organism that comprises either a single cell or a cluster of cells that perform relatively complex functions.)

We've learned a lot…recently the "tree of Life" was modified. Thirty years ago, there were plants, animals and other stuff. Now we have three major domains, bacteria, archaea, and eucarya. And the classification changes are probably not over.

The first domain is for *bacteria*, the best-known prokaryotic cell organisms. The second domain, *archaea*, evolved shortly after bacteria. While archaea consist of single prokaryotic cells, they have very different traits. Not much is known about them; most thrive under anaerobic, extreme thermal, conditions. Oddly shaped mounds of layered sedimentary rock found in Westerns Australia are believed to be the remains of massive communities of archaea. *Eucarya* is the third domain; these organisms have one or more eukaryotic cell (cells with an enclosed nucleus).

Included in eucarya are all plants and animals, sponges, algae, fungi, and more.

Bacteria very rapidly reproduce by cell division which means they have an exponential growth rate. Unchecked, literally within months a particular bacteria specie could dominate all life forms. This doesn't happen because they quickly exhaust the supply of raw materials (food). They then become inactive until there is a change in their environment. They do not die, they simply stop multiplying; they can survive, inactive, for years. Bacteria are routinely recovered after being dormant for hundreds of years.

Thus, bacteria spend most of their time being hungry and doing nothing. However, a change in conditions will likely restart replication and adaptation. Their high adaptability comes from several factors, a relatively high mutant formation rate, and another phenomenon unique to bacteria, a shared gene pool. Under some conditions, bacteria can pass genetic material through their cell membranes from one bacteria specie to another.

Every human has about 100 times as many bacteria cells as human cells. Because we are big and clever and have social lives, we tend to banish bacteria to the fringes of existence. Not so, they were here first and will be here when the sun goes dark...we won't. They got along fine without us; we can't live without them. Bacteria exist under almost all imaginable conditions, from boiling mud pots to ice glaciers, inside rocks, at the bottom of the sea, and deep within the earth. And now probably on the moon. Most are immune to radiation.

Most bacteria are beneficial. However, viruses[3] (and some bacteria) are a major threat to our survival. They come in a variety of forms; some have horrifying effects on complex organisms like us. They are the number three killer of man in the world. Some don't kill us, they simply make us miserable with sneezing, vomiting, and diarrhea. By making their host unwell, the microbe spreads to others, fulfilling their imperative to propagate. (Killing the host might not achieve this!)

Microbes (bacteria and viruses) use several different strategies to spread, one of the most effective is to utilize a mobile third party such as the mosquito. Mosquito-born infections are some of the worst: malaria, yellow fever, encephalitis, etc. We are fortunate that HIV, the AIDS agent, does not spread in this way. The bird flu virus is also a killer; fortunately, it does not readily spread to humans. At any time, we are just a few mutations away from another killer virus that devastates humans. The Covid-19 virus is a killer that threatens all of us. It seems likely it was created by man under laboratory conditions. We might say only man could create such a menace...but that would be wrong. Many deadly viruses evolved without man's assistance.

Another insidious trait of some virus is that they can go dormant for many years and reappear with serious consequences. One of the best-known examples of this is the virus Varcella Zoster. It first causes chicken pox, usually in young children and then hides out in nerve cells. Many years later the virus can reemerge resulting in Shingles, usually in older people.

---

[3] A virus is a Prokaryote cell that cannot live outside the confines of another cell.

Complex organisms evolve immune systems to protect them from microbes. Our immune systems learn new defenses when faced with new forms of virus. Some of our immune system reactions to a microbe attack are so potent that they leave us weak or even dead. (An evolutionary overshoot.) In fact, there is an ongoing battle, microbes adapting to penetrate our immune system and our system adapting to defend against new varieties of microbes. We occasionally lose; it's called an epidemic or a plague. Over the history of civilized man, microbes have played a larger role than guns in shaping the winners and losers.

> *Microbes exist today and will exist as long as there is life on earth. Most are beneficial; without them complex life could not exist. Some microbes are an ongoing threat to our well being and under some conditions can threaten an entire civilization.*

The Wondrous (Eukaryote) Cell

By 2 bya, oxygen levels elevated from 5 to 20 percent of current levels. Uniquely, our planet then appeared blue with massive oceans and a clear sky…bacteria (prokaryote cells) were the dominant life form. About that time a new species appears, a single celled organism with a eukaryote cell. Eukaryote cells are much larger and more complex than prokaryote cells; however, their distinguishing characteristic is that the cell's DNA is contained within the cell nucleus. The nucleus is just one of many organelles (blobs of activity) that perform specific functions necessary to produce the many different proteins needed by the cell.

The eukaryote cell is the building block for all complex life forms and represents a major step up the complexity of life. The immediate payoff for the new complexity is that it could survive and prosper in a world populated with bacteria. Could complex life evolve without these cells? Probably but it would take a different innovation that simply did not happen here. Each cell is a remarkable factory which 1) generates the energy it needs, 2) absorbs and then moves material from within the cell, 3) assembles organic material into proteins for cell maintenance and cell division, and 4) eliminates unneeded by-products. Every activity of the cell is under the careful control of specific enzymes.

Most scientists believe that the original eukaryote cell formed by different types of bacteria merging and recombining with different functions in separate organelles, all enclosed within a single membrane. While scientists have ideas, they cannot define a pathway. Some argue that irreducible complexities block evolution of the eukaryote cell. This argument is used when we cannot visualize how something could possibly evolve. What one must note is that the number of irreducible complexities shrinks as time and man's knowledge and imagination expands. The evolution of these complex cells would have been an extraordinarily complex and difficult pathway. However, given time, complex and wondrous things happen, even miracles.

Various types of algae evolved from the first eukaryote cells, which gave the oceans a tint of green color. Another branch produced fungi, slime molds, and other simple animals, a few of which exist today. In our first biology class, we were amazed as we looked at these amazing little critters with a simple microscope. With a powerful microscope we see something even more amazing. Within the eukaryote cell there is a beehive of activity with tiny particles shooting all over the place. Small bits of DNA are copied into what is called RNA messages and shipped to a ribosome. There, new enzymes are made according to the information contained in the RNA message. These enzymes catalyze the production of new proteins. Raw materials flow through the membrane walls and are transported to various organelles where they are consumed. Waste products are trucked away and exhausted back through the cell membrane. Within the nucleus, DNA splits, combines with amino acids, and becomes two DNA strands with identical information; the nucleus then divides. ATP, the energy source for all cell operations is manufactured in a mitochondria organelle and stored and distributed as needed to the other organelles. And all this happens at a dizzying pace. These are truly wondrous little single celled creatures.

The eukaryote cell has many capabilities that bacteria do not have. First the isolated DNA provide for a more stable inheritance. The greater complexity of the cell provides for well developed, sometime complex, traits to deal with its environment. Most of the early forms were algae which exhausted even more oxygen than cyanobacteria. These animal precursors have energy efficiencies about ten times higher than most bacteria.

The eukaryote cell, the building block of all complex life forms, was a remarkably successful innovation. The evolutionary pathway from bacteria is pretty much a mystery. The miracle of time satisfies most; a few believe that an intervention must have occurred.

Single cell versions of the eukaryote cell multiplied by cell division; they moved to new environments, mutated, and adapted. They were the ancestors of present day single celled amoeba and algae. Together with the many forms of bacteria they were the only life on earth for the first 3 billion years.

The Survival Machines

About a billion years ago a major innovation appeared: the first organism with **specialized** cells that performed special functions. The first to appear were "skin" cells on the exterior of a cluster of cells; these cells served to provide a protective cover for the entire cluster and enhanced survivability of the aggregate. When a skin cell divides it produces more skin cells. There is evidence of these first multi-celled organisms from as early as 1.2 bya.

Over a long period of time, cells within the cluster evolved other specialized functions, again enhancing the survivability of what now is a "multi-cell" organism. However, all specialized cells have a characteristic which represents a shift in life's strategy. When a multi-celled organism replicates, the first cells (now called stem cells) replicate without specialization. After a specific number of replications, specialized cells began to form in addition to the stem cells. Both types of cells continue replication growing the multi-celled organism. The nucleus of the stem cells contains all the information necessary to build the entire organism, and these cells can replicate indefinitely. The specialized cells will only replicate a fixed number of times and therefore have limited lifetimes. In the process the DNA of the specialized cells may be altered or damaged by the environment. However, the damaged cells die with the organism's body and the damage is not passed to new generations. This is a good thing for evolution; many lifetime changes degrade the cell structure and would therefore degrade future generations. Fortunately, these cells and their DNA are not needed for future generations. Only stem cells contain the DNA necessary to replicate the organism.

Thus, a multi-celled organism has a finite life; it acts as a vehicle for propagating the genetic material of the stem cells. They are survival machines[4]...for the organisms' genetic material.

The purpose of stem cells is to pass undamaged DNA to the next generation. This is done via sexual reproduction. This major innovation occurred at the time of or shortly after the appearance of multicelled organizamis. Even in simple organisms it is a two-step process. First comes meiosis which is the halving of the DNA to form a special cell called a gamete (or sex cells). This is followed by fertilization which is the fusing of two gametes with different DNA resulting in a fertilized cell. Scientists have few clues how this might

---

[4] Some would say that a chicken egg is the chicken's way of making another chicken. Modern evolutionist would say the chicken is the egg's way of making another egg. Their view clearly deflates the significance of what we view as our persona.

have evolved; however, it is closely tied to the development of specialized cells.

Sexual reproduction is the last innovation we will discuss in life's survival plan. It is much slower than reproduction by simple cell division, and therefore would not seem to be favored by natural selection. The reason it prospers is that it shuffles the genetic material inherited by the offspring, providing it with modified traits and therefore imporved adaptability. Importantly, these offspring are potentially all viable. Sexual reproduction gives, on an ongoing basis, offspring with new traits with proven viability. Cell division also produces variation in the offspring, but it occurs by mutation of the DNA. Howerver this variation is random and most varients are non-viable. Mutation plays a much lesser role in sexual reproduction in that damage in one gamete can be masked by DNA from the other. Only rare occasions when a damaged gametes from both parents are paired, are the offspring damaged.

Thus, we see that, like bacteria, multi-celled sexual organisms have a rich diverstiy of traits that allow them to readily adapt. However, bacteria indescrimitely swap DNA and each new version has a low probablity of being viable. Their rapid replication gives high number of survivors but an even higher number of causalties. Sexual reproduction produces a fewer number of offspring with diverse traits, but most of them are viable. Thus, species that sexually reproduce have adaptability without the waste and carnage that occurs with more primitive cells that reproduce by cell division.

Sexual reproduction is the primary method of reproduction for most visible organisms, including essentially all plants and animals. The model then for complex organisms is sexual reproduction resulting in offspring with DNA from both parents. The organism lives, competes, reproduces, and then dies; the genetic information survives when their offspring survives. Organisms that perform these steps well in their environment survive in greater numbers (specie prospers) than those that do not. This is the evolutionary process guided by natural selection.

> *Over millions of years, many new organisms emerged, some more complex than others. In multi-celled organisms, what we think of as the organism dies and only its genetic material survives in its offspring. Together with sexual reproduction, this new life strategy is found in essentially all plants and animals. The organisms compete to live and reproduce and then die with only its genes carried into the next generation. The organism is sometimes called a survival machine...survival for the genes.*

Robust Life

By 800 mya all the major innovations needed for complex life have occurred. And, after 80% of life's history on earth, there was visible evidence of something happening. There is lots of green color, in addition to the blue oceans, and some lumpy organic matter here and there. Oxygen levels have risen from near zero to about 20% of current levels. The production of oxygen from photosynthesis continues to exceed the combined consumption by animal respiration and oxidation of natural materials; the percent oxygen in the air slowly increases.

Life requires water, a few basic materials, some fancy organic chemistry, and a source of energy. About 4 bya earth had the essential requirements. There are likely billions of other places in the universe that have similar conditions. Have packets of organic chemicals fidgeted to life in these other places? Very likely. Did cyanobacteria or something like it come along and transform the environment to something suitable for complex life. Maybe. Did something like the eukaryote cell evolve with efficient energy metabolism? Less likely, but perhaps something like it. Did this complex cell develop specialized cells, sexual reproduction, and survival machines? Who knows? The likelihood of life is high; the likelihood of complex life like ours is much lower.

However, a more interesting question is whether there are organisms out there that in any way resemble what we view as intelligent life and whether their existence overlaps the extraordinary narrow window of our consciousness. Would we view our world of from 3.6 bya to 800 mya interesting? Probably not. *An observer like us would likely consider our planet dead for 80 percent of the time that life existed on earth.* Further we have only had the capability to ask these questions for at most one century out of the 37,000 centuries that life existed on earth. If other life exists, it is not likely we would have found it unless it flourished in the same time window as ours. This further reduces the likelihood that it will be discovered…if it exists.

In the following 800 my, two factors shaped the evolution of life. The first is, decent with modification, the mechanism that has been at play since the appearance of the earliest replicator. This is essentially Darwinism. It is not a theory; it is a how living systems adapt to changing environmental conditions over time. It is the survival of the fittest. It has been understood for almost 150 years and still not entirely accepted.

Only recently have we understood the second factor, extinction events. When these events occur, they overwhelm the evolutionary process. They destroy many of the species and restart evolution in a new place with new winners and losers.

> *Living organisms counter, at least locally, the slow dispersion of mass and energy throughout the universe. For this reason alone, they are special…it is the only force in the universe that makes a difference. This would be true regardless of whether or not we existed.*

CHAPTER 2

Life Come and Goes

Life appeared on earth and slowly evolved for the past 3 billion years. One billion years ago it was mostly bacteria plus some simple multi-celled species. It proved to be robust, surviving massive environmental challenges. Biological adaptation enabled some organisms to survive and prosper relatively to others. Until recently, survival and adaptation by natural selections was considered the major factor controlling the evolution of life forms. Now we know extinction events also play a significant role.

Extinction Events

An extinction event threatens organisms that are living at the time; they give no opportunity for adaptation. In some ways, they are like lotteries that drive some species to extinction and allow others to prosper in their place. After an extinction event, there is a rapid biological radiation by the survivors, filling gaps. However, many of the newly radiated species are not competitive and do not survive.

The first great extinction event (that we are sure of) was Snowball Earth, some 700 million years ago (mya). The last was the KT extinction event some 65 mya. The most recent ice age that began a million years ago (ended about 100,000 years ago) is merely background noise. It would not be noted if it had not impacted hominid[5] species.

Snowball Earth

About a billion years ago, Rodina, the first super continent, began to form. It was a hostile place without life of any kind. Over hundreds of millions of years, it had a dramatic effect on climatic conditions and ocean currents. Many believe it was the trigger for the first extinction event. From 750 mya until 600 mya, the earth suffered the most severe ice age in its history with glaciers extending to the equator and ice sheets up to a mile deep near the poles. This event is called snowball earth; it defines what is now called the cryogenic period.

All living organisms were devastated except in a few remaining warm areas deep within the oceans. Since we hardly know what life existed at the time, we know little of what species were destroyed. We can be sure that bacteria and some early multi-celled organisms survived; we know that because they existed before the event and they are still with us.

Our earth has experienced many glaciations; the polar ice caps have come and gone many times. As an ice cap extends from the pols, the earth reflects more sun light, and the earth is therefore likely to get even colder. This reinforcing feedback mechanism can drive glaciations to

---

[5] Hominid Specie is defined as a member of the primate family of which *Homo sapiens* is the only extant species.

extremes. It reverses when the insulating effect of the ice raises the earth's surface temperature and melts ice faster than it is formed. Or when carbon dioxide (or other "greenhouse gasses") reach high enough levels to create a greenhouse effect, and a warming trend. These warming trends can be relatively rapid.

From 650-545 mya, the oceans warmed, and new life evolved. There is a small but growing record of life from this period. Some new and highly diverse body structures appeared; most did not survive. These multi-celled creatures were small, soft bodied, with limited functions. They spent most of their time laying in the sediment. Some developed a top and bottom, front and rear, mouth, and anus. In general, they did little but eat, eliminate waste, and reproduce. Jellyfish and sea worms are descendants of these early creatures. Some grew leaving imprints that are visible to the naked eye. More and more evidence of larger plants and animals are appearing, primarily due to more sophisticate examination methods. There is some evidence of early-stage arthropods (bugs) and mollusks; however, paleontologists are still searching for the emergence of the first bilateral organism (an organism having two sides with internal organs).

A jellyfish has no real brain; it does have neural connections that coordinate its movement. The absence of a brain is not an impediment in many environments. In today's world, jellyfish are rapidly expanding in numbers and size and are migrating into essentially all temperate waters; in fact, they are slowly "polluting" the world's oceans, and man is hard pressed to stop them.

It is believed that some forms of flat worms from this

period developed a nerve center (early brain) capable of transforming multiple stimuli input into coordinated action, like attacking another worm. Some investigators claim evidence of memory in an organism as early as 700 mya. And at some marvelous time, an organism accumulated a set of genes that said, 'make eyes.' This was a competitive winner and eyes were soon found in many animal species. The eye evolved to give the organism accurate readings on things of consequence to that species. Therefore, the eye of every species differs in function, capability, and complexity. Why don't plants have eyes? There is a cost to essentially every inherited trait; the organism must devote resources to it in lieu of others. If that trait does not have a comparable competitive value, it will not be encoded in the genes. Plants don't have eyes because they have little use for them.

> *The first extinction event was a massive glaciation that brought sea ice to the equator. Most life went extinct, with only bacteria and simple multi-celled life forms surviving. As the earth warmed, new life forms evolved from the surviving species. They were small, soft body creature with limited function; most of these life forms no longer exist. However, some evolved important traits encoded in their DNA which were inherited by future generations. Our ability to obtain solid evidence of these early organisms, is remarkable, and a tribute to the work of marine paleontologist. Fossils (calcified bone) won't appear for another 50 million years.*

Life's Big Bang

The Cambrian period began 545 mya. About that time, there was a huge increase in number and diversity of life forms. Hundreds of new phyla "suddenly" appeared. (Phyla are groups of species with common inherited traits.) Some species were very strange with body plans unlike anything seen before or since. One had 5 eyes; another had claws at the end of its snout. Another was disk shaped like a slice of pineapple. Most of the new species did not survive. Evolution always produces variations in excess of what will survive.

We know of this Pre Cambrian biological expansion because, finally, we have real fossil evidence: living organisms from this period had sufficient structure for their remains to be embedded and preserved in the earth's crust. However, it takes very special conditions for fossils to form and the ones that do are not necessarily representative of the period. And, of course, the absence of a fossil record proves next to nothing. Modern day paleontologists take special care in understanding fossil records. In Earlier years, many fossil discoveries led to erroneous conclusions.

In recent years, there have been some second thoughts on the number of new phyla that appeared during the pre-Cambrian expansion. When potential new life forms are examined, their differences from known forms capture our attention. Large differences would seem to imply a different species. An alternate, preferred, way is to classify animals by their common traits. When this is done the unusual features have a less dominant role. A good example is an elephant and a shrew. Elephants are huge, have a really unusual trunk, and they chew on trees. Shrews are small have a little nose with whiskers, and snack on flies. However, both have 4 legs and 5 toes, are warm blooded, have mammary glands, and give birth to live offspring. Both belong to the same sub phylum, mammals. When common traits are used to classify the animal types that appeared during the Pre Cambrian expansion, a plausible taxonomy[6] evolved with many fewer phyla. Short term (again in geological terms) adaptations explain the novelty and diversity within a specie.

To be sure, there was an explosion in the number of fossils just proceeding the Cambrian period. And the biology was distinctly different from earlier periods. However, it seems likely that many of these new life forms actually evolved earlier and only gained sufficient size and structure to leave fossil remains during this period. A better understanding of the Cambrian explosion is that during that period many life forms evolved out of infancy and popped into adolescents. One further point, the "explosion" of new life forms occurred over a 30–50-million-year period; not exactly dynamite.

There are plausible explanations for this expansion in the number and types of life forms. The most prominent being the increase in oxygen levels that occurred just prior to the Cambrian period. Another factor has just recently been understood. Following an extinction event, such as snowball earth, permissive ecological conditions exist where life forms expand into vacated niches. These new species face

---

[6] Taxonomy is classification of organisms in an ordered system that indicates natural relationships based on inheritance. It is frequently modified due to new discoveries and/or a change in bias of those doing the classification.

fewer predators than normal. More diversity evolves than can be sustained as the more competitive species develop and displace others. (A pineapple ring just can't be too competitive!)

If the above arguments seem overdrawn, it is because some believe the pre-Cambrian explosion is evidence of a divine intervention. Massive new genetic information was needed to initiate the development of the new life forms of the period. A possible explanation for this expansion is a divine intervention. However, as more and more data evolve and more is known, this explanation seems unnecessary.

The most significant event of the Cambrian period was that many animals learned to secrete skeletons. Arthropods, the largest animal phylum then as now, evolved a "skeleton" on the outside of their bodies, which served as a shield or armor. Crabs, scorpions, spiders, and insects are all arthropods. About the same time a small worm like creature with a primitive spinal column, became the ancestor of all vertebrates. Soon, a fish species evolved with an internal skeleton; it is called a bony fish (dah). This innovation gave its owner a firm structure that grew appropriate to the size of the organism. Over time, the internal skeleton enabled much larger body sizes. Most large animals that exist today are vertebrates.

> *Marine life developed significant diversity in a relatively short period of time. Two animal phyla evolved skeletons, arthropods, (the ancestor of all bugs,) and the bony fish, the ancestor of all vertebrates. There are abundant fossil remains from this period providing grist*

> *for the taxonomists. And their work is never done; many of the new fossil discoveries raises new issues.*

Trilobites and Friends

Charles Doolittle Walcott is known for the discovery of well-preserved fossils in what is known as the Burgess Shale Site. His story is uniquely American. He quit high school to follow his passion for nature and the collection of fossils. His work in the Burgess Shale and other rich fossil sites transformed our view of early marine life of the Cambrian period. By the end of his career, he served as a Director of the US Geological Survey, head of the Smithsonian Institution, and advisor to then-president, Theodore Roosevelt. A large percentage of 65,000 fossils he recovered and described were trilobites.

One of the most famous arthropod sub-phyla was the trilobites; they hold the record of being the animal that survived for the longest time. They have quite a story to tell.

There were over 1500 species of trilobites. Some were as small as water borne fireflies, others as large as lobsters. They represent more than half of the known fossils from the Cambrian period. They get their name from 3 distinct body parts, head, thorax, and tail: many look like beetles with way too many legs. Like other arthropods, they have exoskeletons. Many of the species were small animals that filtered mud to obtain food. Others were active predators. Compared to anything that preceded them, they were extraordinarily capable; both complex, and adaptable.

Trilobites would have observed some of the pre-Cambrian

explosion. They existed for over 300 million years. They saw the land move; on two occasions, a single land mass formed and then broke apart. Trilobites were at their peak and ruled the seas about 480 mya. About 470 mya they would have observed the first life to appear on land. Primitive plants, perhaps evolved from green algae, appeared along the edges of oceans and lakes. Over time small shrub-like plants appeared.

At the end of the Ordovician period, about 440 mya, an extinction event was rough on marine life. There is evidence that it took the form of several sequential glaciations. (Some speculate that the glaciations resulted from a massive dose of gamma radiation emitted from a distant super nova explosion.) Even the hearty trilobites took a hit… but survived. The extinction cleared the oceans of many immobile filter feeders and created improved conditions for darting fish. We know this from fossil remains. Likewise, some marine arthropods (trilobite cousins) became very large and more aggressive.

Small arthropod (bug) species likely ventured onto land since the early Cambrian period. However, radiation from the sun would quickly destroy them. As the oxygen level in atmosphere went up, an ozone (a reactive form of oxygen) layer formed, surrounding the earth, providing UV protection from the sun. With less radiation, arthropods survived on land and in a short time there were massive numbers of spiders and insects on land, living off microorganisms, plants, and each other.

Fish were making advances as well; their brain structure evolved in a way that gave them the ability to return to their spawning grounds. For the first time known to us, an animal had a memory; genetic coding for memory is a clear winner. We also know that the first vertebrate sub phylum on land, the tetrapod's (vertebrate with four legs), came ashore about 370 mya. They were amphibians; some speculate that on occasions, fish with a memory found themselves briefly on land as they attempted to return to their spawning ground. Some of them spent more and more time on land and became the first amphibians. Initially, amphibians were small but overtime some evolved to over 12 feet long.

Another extinction event occurred that ended the Devonian period (360 mya). Not much is known about this event; land species, both plant and animal took large losses, but most species survived albeit in much smaller numbers. Same for the trilobites.

Land vegetation continued to prosper. Tall trees needed innovations to move water from ground level far into the air and to redesign the placement of leaves to capture the sun's rays for photosynthesis. Near the ground, vegetation was very thick, much like today's tropical rain forest. The primary trees of this period are now extinct. The abundance of vegetation left an accumulation of dead organic matter that did not completely rot (oxidize). Over time, a minimum of 100 million years, these accumulations were compressed and formed solid fossil fuels such as shale and coal. Thus, the name of the period, Carboniferous.

When a large amount of carbon is buried without being oxidized, the oxygen level in the atmosphere increases, this time to current levels. The ozone layer became more robust, further reducing the deadly effects of UV radiation. The thick forests and oxygen enriched atmosphere were good for the arthropods. Extreme varieties included spiders as big as a man's head, dragon flies as large as eagles, and 10-foot-long millipedes that bite. For millions of years some of these monster insects and spiders ruled. However, since they are only bugs, they were not classified as a dominant animal kingdom during this period.

The transition from an aquatic fish to an air-breathing amphibian was a big deal in evolutionary history. It required a modified body plan, sturdier limbs, and a stronger backbone. Also, a rudimentary ear evolved and a more complex lung for breathing oxygen from the air as opposed to obtaining it from dissolved oxygen in the water. These innovations provided amphibians with enhanced land mobility. However, their competition on land included some very nasty arthropods which had been there for millions of years. It must have been quite a battle, early vertebrates (walking, breathing fish) against the bugs.

And where were the trilobites? We can only assume that they attempted to migrate to land and failed. One wonders how things might have turned out if trilobites were able to adapt on land as well as they did in the oceans. They surely would have kicked some amphibian butt. Nevertheless, they remain the darlings of the paleontologist due to their diversity and the abundance of fossils.

The amphibians won the battle with the bugs and prospered; they became the first animal dynasty to dominate land. After millions of years, a new sub phylum broke off from the amphibians. This animal group, the synapsips, is commonly known as the "mammal-like" reptiles. Since mammals didn't exist at the time, and they had the appearance of small reptiles, their name was obviously assigned with bias, probably by a mammal. They are important because they are the ancestor for reptiles, dinosaurs, birds and, you guessed it, you and I.

A relatively minor extinction event crippled the amphibians and the synapsips became the next animal dynasty to dominate land, gradually displacing the amphibians. One of the synapsid species grew to over 10 feet long and developed a sail like structures on its back. The dorsal sail helped moderate their body temperature and allowed them to adapt to more diverse climates. Unlike amphibians, they laid eggs on land and protected them; however, they did not nurture their young. In fact, they often ate them. (Poor parenting habits.) They showed no social behavior; predators could pick them off one by one. They are often confused with dinosaurs which had not yet evolved.

Most fish remained in the sea and prospered. Some were mean and gigantic, making modern-day sharks look benign. The trilobites were more often becoming prey. If the trilobites paid attention, they would have noticed some historically significant branching on land about 300 mya. Reptiles, mammals, and birds split from the early mammal-like reptile. They may have missed the event because none of the new species would be terribly significant for another 150 million years.

Animal brains evolve to meet the challenges of their specific environment. An intelligent animal is one that is well adapted. Some animals evolved with the capability to give some events a fast responses and to insulate them from others. All animals experienced only a small fraction of the outside events; if it were otherwise, they would be totally unable to make sense of the input. In general, a species experiences only the slice of their world that is important for their survival. And mental capability develops that best serve the species survival needs at the time it evolves. Therefore, an animal's brain functions correlate with its environmental challenges. These brain functions, as with all other physical traits, are captured in the animal's genetic inheritance and are passed from generation to generation. This was true for ancient life forms and is true today.

By 300 mya we see signs of patterned (instinctive) behavior which is of high value to future generations. Some bony fish would school to capture more food and provide a collective defense. Ancestors of the spiny lobster would follow each other to an improved environment, a key requirement for their survival.

For over 300 million years the trilobites saw a major expansion in life forms, from the beginning of the Cambrian period to the end the Permium, despite (or because of) two major extinction events and several minor ones. The seas teemed with plant and animal life. Plant life came ashore, radiated, and produced massive amounts of vegetation. When it died, it was crushed and buried, and later converted to coal, the most abundant fossil energy source on the planet. Bugs came ashore and thrived followed by the amphibians which feasted on the bugs. By 250 mya amphibians gave way to the synapsis, the common ancestor of reptiles, mammals, and birds. However, a monster extinction event is about to shuffle the deck; unfortunately, it will knock off our tour guide, the trilobites.

The Permian Slaughter

Permian is a period of geologic time from about 300 to 250 mya. The whopper of all extinction events, the Permian Extinction, occurred 250 mya. It wiped out over 90 percent of all land animals and 95% of marine life. It was the closest life came to total extinction. It was a sudden event, geologically speaking; however, it may have taken a million years or so to fully evolve. It ended the synapsis (mammal-like-reptile) animal dynasty, and the trilobites vanished. Essentially all marine species were staggered. Fortunately, a few reptiles and mammals survived but their time had not yet come.

The causal event is believed to be a period of volcanic activity[7] characterized by massive amounts of lava flowing from huge cracks in the earth's surface. This activity, called a flood basalt, covered a land mass about the size of continental US. For hundreds of thousands of years, lava oozed from the ruptured earth's surface. A flood basalt is rare. Minor ones occur about every 20 million years. The one causing the Permian extinction is the largest known. If it were to occur again, there would be another mass extinction with about the same disastrous results.

---

[7] There is some evidence that the volcanic activity occurred because of a collision with a large asteroid; however, a suitable impact area has not been found.

The lava flows discharged huge amounts of sulfur dioxide, carbon dioxide, and chlorine, poisoning the atmosphere. The rain that fell was massively acidic. Rotting organic material chews up oxygen, further raising carbon dioxide concentrations resulting in a greenhouse effect. The earth's temperature increased by approximately 15 degrees. A drought dominated the planet and land became barren. Low water, low oxygen levels, low food supply, and deadly air resulted in an almost complete wipe out of land animals. Low oxygen levels in the sea destroyed plankton which is the first step in the marine food chain. Most sea life went extinct, and some truly nasty bacteria evolved that thrived in the very hostile conditions.

The devastation that occurred is well documented; however, the cause is speculative; and other volcanic events have not had this effect. Perhaps the size of the lava flows, and the extended period of activity was a factor, or some other unlucky coincidence. What we know about this event has been uncovered within the last 25 years. We are bound to learn more.

Things were grim for millions of years following the extinction. It was 20 million years (twice normal time) before small to mid-sized animals again roamed on land, and even then, with limited diversity. For the first time conifers emerged replacing the few surviving ferns as the major plant type. Oxygen levels recovered.

By 220 mya, a small reptile like specie emerged; its preferred food was insects. The innovation that set the species apart was a complex joint structure in its rear legs. This gave the specie bipedal capability and greater agility and mobility. By 200 mya the earth had regained its diversity. The small insect eating species was the ancestor of the dinosaurs which existed on planet for 150 million years, second in longevity only to the trilobites. The dinosaurs became the dominant animal kingdom. By 150 mya some dinosaurs had grown large; mammals remained small. The dinosaur body plan included hollow bones with air sacks enabling highly efficient use of oxygen. Their large size meant they had voracious appetites.

Dinosaurs generally competed among themselves, and most were both predator and prey. Their life consisted of hunting, fighting for available food and territory, and reproducing. Dinosaurs adapted to a wide variety of environmental conditions. A herbivore species grew to become gentle giants that munched on treetops. A carnivore species became the finest killing machine the world has ever seen.

Mammals also developed internally with placenta, an expanded brain and, for the first time, a cerebral cortex. In time this brain addition will support social behavior, language, and problem solving. A brain is a high energy consumer and requires the animal have a high metabolic rate; it is likely there was a tradeoff between more complex brain and the overall size of the mammal. Being nocturnal, they developed a keen sense of hearing; this required a sophisticated inner ear and additional brain structures.

Compared to reptiles, mammals of the time reproduced in large numbers and were short lived…this resulted in more adaptability. Other innovations were fur, warm blood, ability to nurture eggs inside their body, and a portable supply of milk to nourish their young. These innovations required additions to their brains to host their many new emotions.

About 100 mya the first known flowers appeared. They became symbiotic with insects. The insects pollinated the plants and grew fat on plant nutrients such as honey found in flowers. Both the insects and flowers were enhanced, and diversity was stimulated. The new food source for insects resulted in new varieties of insects with higher protein content. Flowers also benefited mammals; they began to eat seeds that provided further nourishment.

> *The Permian extinction took a heavy toll on all life forms, plants and both marine and land-based animals. Surviving vertebrates included reptiles, mammals, birds, and dinosaurs. Dinosaurs prospered becoming the dominant animal kingdom. Mammals remained small and nocturnal...however their brains developed with even more capability.*

A huge asteroid struck the earth about 66 mya, resulting in what is known as the KT extinction event[8]. While not the largest, it was the last and best-known major extinction event. When the event was first proposed in 1980 it was greeted with skepticism. For one thing, it was proposed by physicists based on trace amounts of iridium found in fossil beds. Paleontologists, who had studied these fossils for hundreds of years, had difficulty accepting the proposal

---

[8] The name KT indicates the boundary between the Cretaceous and Triassic periods. (Some folks spell Cretaceous with a K).

from outsiders. It was finally confirmed in 1999 when the caldera was found off the coast of Mexico.

The impact of the asteroid created a super-heated blast wave that carried for thousands of miles in all directions killing all in its wake. Tsunamis up to one-thousand-foot-high flooded coast lines around the world. The dust entering the atmosphere resulted in years of little or no sunshine; the rainfall was highly acidic. Most of the earth's plant life died leaving little food for land animals. Dinosaurs were among the first to die along with most other large animals. Marine life was also compromised. Smaller animals including a small, warm blooded, nocturnal animal survived. On came the mammals.

Mammals have several survival advantages over dinosaurs. However, they are of little use if the mammal is eaten whenever spotted by a dinosaur...which had very good vision. With a major extinction event, biological adaptability is not a factor. Mammals survived because they were small and could find sufficient food to survive.

> *Dinosaurs were destroyed by a cosmic event, a large asteroid collided with earth. It was not the first such collision nor the largest. Without the KT event, mammals would not have prospered. They may have prospered later, or they may have gone extinct. The earth being hit by an asteroid was lucky for us, not so for most other species.*

Mammals Rule (Finally)

From 65 mya onward, mammals prospered. They grew in size and diversity, and migrated over the entire world…over time they became the dominant land animal. Some returned to the sea and are the ancestor of whales and dolphins.

Why did they prosper? Initially, because they were no longer dinner for dinosaurs. Mammals have some important innovations; fur, warm blood, ability to nurture eggs within their bodies, and a portable supply of milk to nourish their young. And, relative to reptiles, they are social animals, meaning that they would often live and hunt in groups. These innovations were accompanied by additions to their brains to host the new emotions needed for nurturing young and to support social behavior.

Many mammals have bifocal imagery with two forward looking eyes. Their brains developed the capability to process these two slightly different images into a 3-dimensional view, a marvelous development with definite survival value. Before long, hoofed mammals with clawed feet evolved. This epoch time saw the development and proliferation of horses, rhinoceroses, pigs, giraffes, sheep, goats, antelopes, and cattle to name a few. Some grew to giant size, like the saber-toothed cats, giant ground sloths, and woolly mammoths. The wide diversity among mammals was expedited by the breakup and movement of major land masses. When this happens the species, now separated, encounters very different environments and therefore evolve different traits.

Divided over time, primates evolved from their mammal ancestors. They are characterized by a refined development of the hands and feet, a shortened snout, and a larger brain.

About 15 mya, primate ancestors of man...gorillas and chimpanzees evolved; they lived in trees in tropical Africa. Both of these ancestors share about 95% of DNA with modern man.

Arguably, mammals and more specifically primates are the most complex organisms to ever exist. A reasonable measure of complexity is the amount of functioning genetic code required for the species to replicate. High complexity implies, simply, that a large amount of genetic material is needed for a species to be what it is.

A word about complexity; it is not a virtue but a consequence of evolution. After each significant environmental event, some species persist that have more complex genetic information than their ancestors. This suggests there is a direction for life's evolutionary process. However, the correct way to view the complexity of an organism is that the added functions were what the organism needed to survive in the presence of other organisms. Bacteria occupy the sweet spot; their simple design is good enough to survive in their environment. New species had to be more complex to survive in the presence of bacteria. And so, it goes up the complexity scale...extinction events do not alter this phenomenon. Primates are complex because they were late to the party. New capability was needed to survive in the presence of the incumbents. Life forms are driven to survive, and complexity pays off only when it makes a competitive difference. When complexity exists which no longer supports survival, it will, over time, be lost.

Thus, the complex traits of primates were not a goal of evolution, but a result of the competition along their torturous evolutionary pathway. Primates have a big brain because they could not now or in the past compete with "lesser" species without it.

> *Mammals finally rule, although only for a short time relative to other evolutionary winners. Their traits have outstanding survival value relative to prior animals that ruled the earth. However, only time will tell; prior dominant animal groups all have a far longer history of survival. Mammals and specifically primates have some very complex traits that are quite special. However, complexity is not a goal of evolution, but rather a by- product of their evolutionary pathway.*

The Information Animals

Some animals have evolved sophisticated nervous systems that enable them to anticipate future events and thereby improve their chances of survival. In primitive animals, all signals from the environment result in a simple reaction and every reaction is essentially immediate. In a slightly more complex animal, the brain enables short range anticipation. For example, a visual signal might cause the animal to duck an incoming rock...before being hit. In essentially all animals these immediate and near immediate reactions are hardwired into their brain structures.

At the next level of complexity, the animal not only anticipates an event, it performs an "analysis" before reacting. For example, when encountering another animal, the brain determines whether it is friend or foe. Or is it food?

Also, complex brains will activate entire patterns of behavior in addition to simple responses. Thus, when encountering another animal, it makes a quick analysis, and then trigger the appropriate behavior; fight, flee, or perhaps, attempt to mate. The analysis avoids some inappropriate responses.

Something important happens at the next level of complexity. It is an advanced behavioral strategy, which only a few animals' practice. The animal will seek additional information for its own use, either now or at some future time. It is a fundamental shift, the birth of curiosity, or epistemic hunger. The animal becomes what might be called informavores. Complex animals, especially primates with their large brains and highly mobile eyes, often use this strategy.

Overtime, the mammals brain evolved into two specialized areas; the dorsal handles the on-line piloting responsibility…to keep the animal out of harm's way. The ventral area has some free time to identify environment hazards and opportunities. And then develop a data acquisition strategy.

Somewhere in the development of primate brains, there was another advance, the evolution of neural plasticity. Neural matter forms links when responding to various events; when these neural linkages are repeated many times, the links become permanent. This is called brain plasticity and it enables 'learned behavior'. The capability for learned behavior is an inheritable trait and it certainly contributes to survivability. The actual learned behavior, however, is not inherited. Brain plasticity is not unique to mammals; however, most other animals have much less capability. The development of brain plasticity is in some ways analogous to natural selection; structures that develop are those that are reinforced by frequent use.

> *Over time, complex brain structures developed that anticipate future events and initiate appropriate protective behavior. Further, some animals began to seek out information that might be of future value. This is a strategic breakthrough with great survival value.*

God of Our (Animal) Ancestors

In an earlier chapter we found that our universe needed an act of creation for life to appear on our small planet Further a very sophisticated and precise set of conditions were required for life to develop into the fullness we now cherish. We say that, even though, some events seem wondrous, even magical; they could have occurred without further help from the creator. Never-the-less, we cannot help but view these developments with reverence.

We have seen how life is about self-replicating organisms, a non-random process controlled by the organism's genetic code. Mutations, which are random events, allow the organism to adapt. Extinction events result in rapid, massive changes. While the causes of the events are largely unknown, their impact on life can only be viewed as large, random, and uncontrollable. Each event resulted in a shift in the biological expansion, with different winners and losers. The combination of adaptation by evolution plus occasional extinction events is the pathway that led to life's diversity and complexity. The pathway was infused with randomness, contains many dead ends, and does not appear to have an end point. It does not look like a divine work to us.

Evolving organisms have a purpose...to survive and reproduce. A successful organism exploits its resources while expanding its numbers...until the resources are exhausted. Then it adapts or dies. Many living species produce thousands of offspring, most die at or shortly after birth. Nature continually innovates, improving her product; however most prior versions do not survive. The delicate balance we see in nature can be more easily characterized as a balance of suffering, rather than harmony. Would a prevailing God favor the most competitive, and tolerate the starvation/extinction of lesser organisms?

On many occasions during the past 3 billion years, life has been stable, only to be destroyed or disrupted by an extinction event. A prevailing God is either wasteful or powerless to control these events. His most successful species, as measured by longevity, has been the trilobites, and they are extinct. His most complex species evolved out of necessity. A god that favors the most competitive, tolerates massive slaughter, and allows extinction that threaten all species, does not seem omniscient, loving or even just. Thus, an attempt to find God in the billion years of life on earth, up to 5 million years ago, does not seem useful.

Summary

There is an evolutionary path from microscopic soft-bodied flatworms to primates, just as there is a pathway from flatworms to cock roaches. In fact, all existing animal species can boast an uninterrupted ancestral pathway from some sort of flatworm. There are many more pathways which dead end. Of all the species that have existed on earth, 99 percent are now extinct. Some species went extinct from a failure to adapt to evolutionary change or a failure to successfully compete. Many more species went extinct as a direct result of an extinction event. If any one of the events did not occur or if one occurred at a different time, life on earth would be very different. If there were a few more extinction events, or if they were more severe, earth might again be void of life.

It is hard for us to accept that we exist because of untimely extinctions and other random events. Five hundred million years of evolution of complex animal life plus 6 major extinction events demonstrate life's abundance, variety, and persistence. However, it also appears to be a crap shoot with most innovations lost to extinction events. Through it all there is a surviving lineage, but not one that has a predictable destiny. In the Lottery of Life, trilobites existed for 350 million years, were highly evolved, and were extraordinarily well adapted to their times. Likewise, dinosaurs lived for 150 million years and were the largest and meanest creatures ever to exist on earth. Both successful life forms were wiped out by external events. Primates appears to be highly adaptable; but stay alert… they've only been around for a relatively short time.

CHAPTER 3

❧

Tribal Man

Life on earth has existed for almost 4 billion years, although there was little visible evidence of this for eighty percent of the time. For the past 500 million years, there has been abundant life. However much of the time; life has been in a restart mode, recovering from massive extinction events. In prior chapters we have raced through geological time to capture life's milestones on earth. And we have covered billions of years. It is staggering to comprehend these time spans; in fact, the human mind is not up to the challenge.

In this chapter we introduce the word **meme**: a kernel of transmittable culture passed to others including offspring. It was popularized by British evolutionary biologist Richard Dawkins and is meant to rhyme with **gene**.

A Primate Stands Up

About 5 million years ago an African primate broke ranks and became a distant ancestor of modern man. While several species evolved from that point, only one survived, Homo Sapiens. We call these new species pre-man species[9].

---

[9] This definition is similar to a hominid species; however, some taxonomists include a few ape species as hominids that are not ancestors of man.

When primates first appeared in Africa, North and South America were separated allowing ocean currents to flow freely between what would become the Pacific and Atlantic oceans. Over millions of years, this separation slowly closed by the formation of what we now know as the Isthmus of Panama. This land bridge had fully emerged about 5 million years ago. As this happened, ocean currents were greatly modified resulting in large changes in weather patterns over the entire planet. One of the many affected areas was tropical Africa which became warmer and dryer. Some jungle areas slowly turned to savannas with fewer trees and higher grasses. This transition took thousands of years and was just one of the many environmental shifts that routinely occur on earth. And it had an enormous effect on the plant and animal life of the region.

Large primates of the time were ideally adapted to jungles; they lived comfortably on the available fruits and plants and had highly developed skills for arboreal life. They did not live as well in savannas. Many other large mammals do, some of them are skilled carnivores. The transition from jungle to savanna was a major challenge for the primates; they were destined to either migrate, adapt, or die. This is not a choice but happenstance. The adaptation pathway was a tortuous route that led to modern man.

One of the first adaptations was that tree-swinging primate increasingly walked on the ground in an upright position. This gave him an advantage in seeing and traversing the more open land. At the same time large predators, lions, tigers, and wolves migrated into the same region. The upright primate had only one defense against these new predators...climb a nearby tree. As the environment continued to change there were fewer trees. Life was very difficult and most large primates did not survive in the savannas. Natural selection favored aggressive male primates with high levels of testosterone. (Running skills also helped.)

Those that did survive roamed the savannas in small groups. They adapted to new foods including new plants and small animals; they became better runners and learned to swim. They scavenged for leftovers and learned to avoid their worst predators. These changes were forced for survival; even then, they were as often prey as predator. The survivors became new species[10], Australopithecus Africanus, the first of many pre-man species. His new upright posture set off many other adaptations.

Fossil remains of this upright primate were found in the Ethiopia in 1974. They were given the name Lucy, after the Beatles song "*Lucy in the Sky with Diamonds*". The story goes that the song was playing in the background when the anthropologists realized what they had found; in recent years many new fossils of pre-man species have appeared, some dating to 6 million years ago (mya). It is believed by some that several species of bipedal primates existed in the time frame 3-6 mya. So far, they have all appeared in Africa and it is assumed they all had the same primate ancestors. Some may well have coexisted for long periods of time. Detailed DNA studies suggest some very early pre-man species reverted to chimpanzee mates for periods of time. Regardless, only one-pre-man species survived this period. And the name Lucy is generally given to that earliest known upright primate ancestor of man.

---

[10] Specie is defined as a group of plants or animals that willingly engage in sexual reproduction.

All primates have emotions that drive them to love and nurture their offspring. Pre-man evolved two additional traits, male-female bonding, and male parental involvement; neither trait is common among other primates. Both enhanced the survivability of their offspring. As pre-man's posture became more upright the female birth canal narrowed. The resulting adaptation was the early birth of offspring. Relative to her primate cousins, pre-man's offspring were necessarily born premature. And the offspring were extremely vulnerable...requiring extraordinary parental care. A trait for male participation in the parenting was therefore favored. While sexual asymmetry[11] between males and females results in different sexual traits, bonding, either in polygamy (one male several females) or monogamy had survival value.

Over time these upright primates formed into small clans (mostly among kin) for mutual protections. Clans grew in size; some clans combined into tribes, resulting in community living with other than close relatives. These larger communities, together with monogamous sexual behavior, represented a significant departure from other primates. New social skills evolved which were needed to prevent the internal destruction of the tribes. Traits such as the exchange of favors among trusted tribe members, the ability to detect cheating, and the tendency to conform and adapt to the wishes and opinion of others...all contributed to tribal cooperation.

As social skills improved, the tribe slowly became a

---

[11] Males seek multiple mates to produce as many offspring as possible. Females, with fewer possible reproductive experiences, seek mates inclined and capable of caring for only their offspring.

cooperative hunter-gather tribe. This transition took millions of years to fully evolve. Before, members of a tribe would scavenge for food acting as individuals with limited sharing. This was displaced by a cooperative hunting effort with sharing of the prey among the hunters in his tribe. Specialized roles developed such as tribal leader and caretakers for the young. Cooperative hunting enhanced food supply: cooperative defenses protected against predators. Groups were not only more efficient, but they could accomplish results unattainable by individual hunters. These lifestyle modifications transformed the character of the tribe. New behavioral traits included a requirement for forward thinking and even simple planning for cooperative hunting.

While harmony characterized the new behavior within the tribe, it was not extended to others. Tribes were territorial. They often raided outside their territory either to broaden their food supply or to capture new mates. All outsiders were enemies, to be killed or raped.

Pre man's up right posture set off another adaptation. Overtime his relatively free hands developed the precision movements needed for tool use. This required more delicate fingers and brain enhancements to control their movement.

Millions of years passed since a few primates became bipedal. This seemingly unimportant event in fact set in motion important changes to the species: mobility on the ground, free hands for tool use, highly vulnerable offspring requiring extraordinary parental investment; coupling, and communal living for added safety and survivability. Their basic social unit was the hunter/gather tribe.

The first pre-man species had competitors for food and a large number of predators. They were often scavengers rather than hunters, prey rather than predator. While some lived in caves, they were incapable of building a shelter. Communication consisted of jesters, grunts and scream. In some ways, they resembled a group of upright, loud-mouthed apes.

Brain On Fire

For social animals, learned behavior can become genetically encoded much more rapidly than by the process of random mutation and natural selection. Take for example hunting skills:

Surviving in a savanna required that pre-man become an (involuntary) herbivore. However, he still craved the high protein treats that his ancestors enjoyed living in tropical forests. Some managed to stand on two legs instead of four and some had agile hands relative to his knuckle dragging relatives. A few had the ability to throw stones at his adversary and his prey. This capability allowed the use of tools for getting more of that good protein.

As some early pre-men began to bring home the treats, they enjoyed newfound fame within their social community. With fame came leadership, authority, standing, and mates. Men capable of selecting the right stones, aiming at a target, and having the coordination of muscle movements to throw a stone and hit a target were the ones rewarded. Those with awkward hunting skills were losers…and had a harder time getting mates.

What resulted was a rapid evolution of men with genetically coded, brain structures enabling superior hunting capability. The fundamental driver for this change was improved survival prospects… more food and more mating opportunities.

This selective breeding, brought about by social behavior within his tribe, resulted in more men with improved hunting skills.

Physical changes to the brain which would enabled the new skills for hunting would have taken hundreds of generations to evolve via mutation/natural selection. With selective breeding they evolved in a much shorter time.[12]

Overall, natural selection with a social overlay had a significant impact on the physical evolution of men in social groups. The change in brain size and complexity is the most rapid evolutionary change that we know in all of nature. While many factors contributed, many were synergistic which accelerated the changes. The innovations to come (cultural transmission, language, and consciousness) all have a profound impact on both survivability and sociability.

---

[12] This same concept occurs when breeding dogs or horses; the difference is that man's social behavior not the dogs that makes the selection.

Only a large complex, highly layered brain could support these new activities.

> *Over a million years or so, the traits that enabled successful hunter/gather living became heritable in the form of innate emotions which were the proximate cause for the behavior. The ultimate cause is genetic material that produces brain structures for these emotions, not unlike brain structures that drive mammals to care for their young.*

> *The very rapid evolution of pre-man's brain capability was synergistic with the development of the more complex social behavior required for successful tribal living.*

Out of Africa-Verse One

The Homo Erectus specie appeared about 1.75 mya. Members of this species are the first that in anyway resemble modern man. His body was much more human-like; his bone structure adapted for supporting bipedal mobility. He had much less body hair and he had sweat glands. He could run long distances at reasonable speeds. His brain supported adaptation to different environments. Importantly, he developed and refined his hand dexterity for tool use. Homo Erectus then had a defense from predators other than climbing a tree; his much-improved competitiveness was driven by his cooperative hunter/gatherer lifestyle.

About 1.6 mya, he invented a hand-held, ax; it was made

by chipping an edge onto a reasonably flat rock. It was very useful for hunting, food preparation, digging, etc, and was the tool of choice for over 1 million years. However, in all that time, a million years is a long time, no pre-man species ever put a handle on it. While his brain was growing in size and complexity, he, in general, lacked innovation.

Homo Erectus was the first pre-man species to successfully migrate out of Africa, across jungles and deserts, to Europe and Asia. He was the first with sufficient adaptability to survive the colder weather and the different food sources available in these new lands. He did this without building a shelter; he still lacked the ability. Much later he migrated to the Far East and the Indonesia Islands, requiring new skills to navigate over water.

About a million years ago Homo Erectus overcame a fear held by essentially all mammals…the fear of fire. He began using fire as a tool. It provided protection from predators when used at the entrance of his caves, kept him warm, and allowed him to tenderize meat. More protein in his diet allowed him to grow taller, and to further expand his brain. He lost the powerful jaws characteristic of most other pre-man species; they were not needed to chew cooked meat. His use of fire is a remarkable innovation…it represents one early example of innovation, requiring imagination and independent action not known in earlier pre-man species.

> *Homo Erectus was the first pre-man specie to resemble modern man. He had a larger brain than his ancestors and brain functions that developed based on several millions of years of tribal living. He still lived in caves, not*

yet capable of building a shelter, or speaking a language. However, his ability to adapt to different environments made him the world's first globetrotter. He is known for tool use, specifically the stone hand ax. Of great significance, he captured and used fire.

Tribal Man and Culture

Pre-man tribes developed distinct cultures. Culture is defined as the information shared and exchanged among members of a community that are routinely transmitted to the next generation. Two essential requirements for a culture are a brain with plasticity that allows learning, and a community that shapes and transmits elements of the culture. Pre-man met both requirements to a greater extent than any other animal species up to that time. His culture began when man first formed into clans for survival. Even without a language he was able to transmit his limited skills and primitive lifestyle.

While genes transmit biological traits, a culture transmits information such as skills, tribal beliefs and tribal behavior. Genetic evolution is governed by natural selection. Cultural evolution is governed by social acceptance by the community that "owns" the culture. A cultural unit is called a meme (rhymes with gene). A meme is any type of activity, behavior, or knowhow that is replicated by the culture. It is replicated when the meme is viewed favorably and worthy of transmission. Just as sexual reproduction and mutations result in a variety of traits, there are interpretations and even mutations to memes. And only the most successful memes survive. This might be thought of as 'evolution by the process of social selection'.

The accumulation of memes for any one social group becomes the group's cultural knowledge. This knowledge can be of extraordinary survival value to the group...or not depending on its veracity. Of course, if the community does not survive neither does the culture. Also, cultural memes can be destructive and even drive a community to extinction.

Cultural evolution and transmission is the second way that communities evolve...and it evolved out of the first. Genetic evolution is very slow and requires generational turn over for change. Early on, cultural evolution was also slow due to the limited cognitive ability of the community. New skills, ideas, beliefs can only be transmitted as fast as individuals in the community can digest and communicate them to others. As skills developed meme repetition became faster and more frequent.

Unlike inherited traits, elements of the culture may be taken and adapted from other communities, although the resistance to memes from outside community is often high.

> *The reason man dominates the planet and gorillas are nearly extinct does not lie in our special DNA. It is because we have learned to accumulate cultural information and transmit it across time and space, potentially pooling the cognitive resources of all men both dead and alive. Culture emerged within individual pre-man tribes and differed from tribe to tribe. The capability to maintain*

a culture and transfer it to offspring is genetically supported.

Man Speaks

Very recently, researchers have identified an enabler for languages in our genetic code. Based on this marker they have estimated that man could speak perhaps a few million years earlier than previously thought. He was capable, but did he speak?

Languages probably developed over thousands, even millions of years, at different times and in different places. Gestures, screams, and grunts became associated with various emotions, objects, and activities. A larynx capable of a rich variety of sounds co-evolved. However, a language requires more than rich sounds and a shared vocabulary. A parrot can learn words by imitation but cannot form a sentence or understand a language.

Our brains consist of many neurons and groups of neurons acting as independent processors; they act out in response to various sensory inputs, or inputs from other neurons. Nowhere in our brain is there a clock that arranges and/or records activity along a timeline. Language requires this; just how it developed, how it works, and the full consequences of it happening are not understood. Like many other primates, pre-man already met one requirement, a love of mimicry. Other requirements are 1) the acceptance of symbolic representations (words) for real things; 2) the concept of linear time (required to sequence words and thoughts), and 3) logic to understand simple syntax. These capabilities are not found in prior pre-man species or any other animal species. Pre-man developed the mental machinery to support these activities and to speak about 150 thousand years ago…about the time one pre-man specie became early man or Homo Sapiens.

Language is the key to our learning and critical for the rapid development of our culture. We now know that language competence requires mental subsystems in many areas of the brain. But how did they develop and how could they develop so fast? Many believe that existing mental systems were re-configured to support language. While this may sound strange, we are learning of another unique characteristic of man's brain…its phenomenal adaptability. This is probably related to the entangling properties of the brains neurons.

Language supercharged the development of tribal cultures. It clarified communications and enhanced sharing of experiences. It greatly enhanced the level of cooperative activity within the tribe which functioned as a group. It also brought socialization to a new level as man could now spend hours sharing stories and building social traditions. Each tribe was building a library of cultural knowledge (memes) that passed from generation to generation. Although each tribal library was different, they had many of the same elements such as their language, tribal history, and information on the use of tools. His mystical beliefs were his religion.

Species knowledge continued to build; Homo Sapiens now have DNA that supports a remarkably complex brain with the ability to learn and use a language. In fact, cultural evolution is a remarkable new pathway for advancement that appears to be unique to man. The new way of learning, cultural transmission, became the dominant feature of man's further development and a survival advantage like no other.

> *About 200,000 years ago a pre man species had the mental capability for a symbolic language; this specie is now called Homo Sapiens (HS). The ability to learn and use a language was unique and universal to HS tribes. Assimilation of this capability was very rapid and was synergistic with increasing brain size. Language provided a massive stimulus to man's culture development. And he retained all his innate traits that evolved over 5 million years of tribal living.*

Man Knows Himself and Is Confused

Another cognitive function evolved that distinguishes man from all other animals. Man became conscious of his own thoughts and feelings, and he learned to express them with words and art. Consciousness meant that to some extent, he knew his own mind, giving him insight into the minds of others. One of man's first use of consciousness was to recognize himself as distinct from all other parts of nature, i.e., he developed an ego. He began to systematically accumulate information about his own experiences, thoughts, and feelings, which became what we know as his "self". His newly discovered self, needed constant reinforcement. Initially, he believed that he was separate from nature; over time some believed they were above it.

Consciousness and the recognition of self-interest was

a gift to man as important as life itself. Without it there could not be independent thought or a basis for individual choice. While the behavior of most animals is driven only by current (or innate) stimuli, man behavior could be driven by thoughts about the past. He could review experiences and form a basis for further actions and even conceptualize events in the future.

We do not understand how and why consciousness first developed. Most cognitive scientists believe that our mind is a product of our evolved, advanced brain and consciousness is a sequential awareness of otherwise chaotic brain functioning. Some view consciousness as a new sensor that overviews brain activity. This implies that consciousness takes place at a specific place in the brain. However, if there is a place in the brain for consciousness, no one has found it, and most do not believe it exists. Some believe consciousness is a necessary byproduct of the sequential ordering of brain functions, a result of imposing linearity on massively parallel brain functions. Some claim consciousness does not exist in any meaningful way and that it serves no new function whatsoever. The relationship between the physical brain, the mind, and consciousness is likely the most complex unsolved problem that scientists face. Recent research with advanced brain imaging tools is helping but progress is slow.

We have encountered mysteries on life's evolutionary pathway; things that have happened that cannot be fully explained. For example, the appearance of the first simple replicator that evolved into bacteria, and the appearance of the eukaryote cell, the building block for all plants and animals. However, the greatest mystery would appear to be the development of consciousness and a concept of self. We are not clear what they are and only guess at how we got them. We argue whether they are real and whether they have a function. And yet every man believes he has both, and, at times, they dominate his thinking. For most, it is easier to believe that these traits are divinely gifted than to accept the natural explanation offered by most cognitive scientists. This is the stuff that religions are for…a fascinating and significant new phenomena with no explanation. For those seeking an explanation, a divine intervention seems to work as well as any other.

Man's consciousness probably evolved incrementally together with language. His newly discovered consciousness led him to place himself in time and space and to reflect on that placement. He now views time as linear, having a beginning and perhaps an end. He became aware that he is mortal and that his life will end before the end of time. He also began to understand that some places might exist even if he did not know of them. Taken together, early man developed a sense of transcendence…a feeling that there are things that exist beyond ordinary experiences.

While some disagree, consciousness would seem to provide some survival benefits. An example would be self-exhortation, self-reminding, self-monitoring that are needed to concentrate on difficult issues. While the role of consciousness in decision making is debated, it is clearly a factor. Regardless of what consciousness may or may not do for man, it is required for freedom of choice and accountability…major issues for civilized man. Those who believe in strict determinism believe that everything we do is simply a consequence of prior events. They do not believe in free choice and therefore must struggle with ethical behavior and accountability. Determinism is not a useful concept in understanding the issues of our time.

> *Our understanding of the evolution of consciousness, its relationship to brain functioning, and the aggregate sense of self is far from complete. Consciousness brought man a new dimension; he now knows his own mind and began to view things from his own perspective and self-interest. He developed a sense of transcendence that raised questions for which he had no answers. He became aware of his own mortality.*

Tribal Religions

Consciousness drives man's need to make sense of his origins and to confront his mortality. Universally, beliefs about these needs and fears became a part of man's ***tribal culture.*** Mystical concepts were often used for closure. The set of ideas held by the tribe about supernatural entities, agency, and future consequences became his ***tribal religion.*** While religious beliefs and rituals differed from tribe to tribe, all tribes had them.

Without religion, man had anxieties, sometimes unbearable, due to issues that were un-resolvable at the time. One of the first signs of that anxiety was the occurrence of burials and rituals about death; the oldest of these occurred about 100,000 years ago. Early tribal religions, even though very diverse, served several important roles. They provided: 1) explanations for concepts he could not understand, 2) an attempt to manipulative forces outside of his control, and 3) justification for some individuals to have power over others. Tribal religions seldom had what we now call moral content.

We can only speculate about tribal religions of 100,000 years ago. However, we can learn from the few remaining tribal communities that exist today. First, the religious teachings are all oral with no written references. Words, shared experiences, and even timely silence communicate in ways not possible in writing. They demand a reliance on memory and are transmitted by the repetitive sharing of stories. Another feature is they are fully integrated into all other aspects of their culture. Also, tribal religions are backward looking, toward what they believed to be the original source of things. (Likely the gods who gave the world order.) The past was their golden period. Ancestors are considered closer to the source and are therefore revered.

Early man viewed himself as part of a tribe and that tribe was embedded in nature; self-identity was relatively weak. The tribe and nature were a part of the same order, like a finger on a hand. There was little separation between their world and the worlds associated with the divine. They assumed the other worlds were there to nurture them and they were not disposed to challenge or escape from them. The concept of "salvation" did not exist.

Notably absent are morality issues. Tribal man lived by his tribal rules. Period.

> *Universally, man adopted supernatural explanations to soothe the anxiety of unanswered questions. Historically religious beliefs have provided explanations.*

> *Symptomatic of religious beliefs, they became*
> *embedded into the culture; they can be an*
> *impediment to change even in the face of new,*
> *uncontradictory information.*

> *While tribal religions different widely, all*
> *tribes had them.*

Behold Adam

About 75,000 years ago, the last super volcano eruption occurred on the island of Toba in Sumatra (Indonesia). This massive event (over a thousand times larger than the Mount St. Helen's eruption) covered Southeast Asia with up to 6 inches of volcanic ash and blotted out the sun for months. For several years, the atmosphere was heavy with sulfur dioxide, and sunlight was limited over the entire planet. It put a squeeze on plants and animals and specifically on pre-man species. Homo Erectus in Asia largely disappeared. Low temperatures and low rainfall killed much of the vegetation, putting stress on African pre-man species and the newly emergent Homo Sapiens (HS). In fact, the HS specie narrowly escaped extinction. The pre-man species in northern Europe (the Neanderthals) were relatively unaffected; they were accustomed to cool climates and environments with limited plant life.

DNA evidence shows that essentially all current mankind are descendants from a single East African Homo Sapiens tribe; we will name the tribal leader Adam. Even with their strong heritage they struggled with years of cooler climate and diminished food supply. But survive they did;

Adam's tribe and their ancestors succeeded where others did not...all other tribes of Homo Sapiens died out. Remarkably, only one clan is the ancestor of modern man; all others failed the evolutionary lottery. Mount Toba narrowed our genetic ancestry to a symbolic Adam.

Once again, the ancestral line between man and the flat worm is impacted by uncontrollable natural phenomena, most of which must be considered random. If Mount Toba had not erupted, other pre-man species would have been viable and perhaps survived into present times. Asia might be populated by different man-like specie, the descendants of Peking man...the dominant species in Asia at the time of the eruption. Peking man, presumably very capable, would have competed with African Homo Sapiens with one or the other surviving but probably not both. Or what if the Mount Toba eruption was more violent and wiped out all Homo Sapiens. This would leave only the Neanderthals who presumably would have survived in the icy northern Europe. Would they develop and populate the earth, or would all hominid species go extinct?

When the climate normalized, the surviving Homo Sapiens thrived; they began to re-populate the African continent. And some showed some novel behavior. It would seem some tribes improved their hunting/gathering skills to the point that they had spare time; for the first time they did some things better than they needed to be done. They also showed some forms of primitive art and used symbols to represent real things. There is evidence of jewelry and clothing as early as 70,000 years ago. This represented the dawn of creativity, driven by early man's new capabilities, language, and consciousness.

Adam, his clan, and their offspring survived in part because of their large highly structured brain... and a whole lot of luck. They developed new cognitive functions: inferential reasoning, abstractions, conscious awareness, self-identity, mental models, complex cause and effect relationships, etc. Nevertheless, most things were done without thought or reason. They were driven by circumstance, chance, and nature; like their ancestors, they seldom made choices. Instinct and survival requirements strongly influenced their actions. Their tribal rules, which might be called tribal morality, evolved over millions of years governed their behavior with respect to their environment and with others within the tribe. They accepted changes in their natural environment as fate. They could not visualize let alone evaluate most alternatives.

Another missing element was agency; man did not view himself as the driver for his own actions and was generally unable to accept responsibility for outcomes. Without agency, there is no free will. However, a few men were beginning to act out of their own self-interest; to feel special, not just separate from nature, but in control of it, and willing to do things contrary to their tribal leanings.

About 50,000 years ago the last great ice age gave way to warming. Moderate climates brought abundant food sources. Many mammals died out during the ice age; early man survived and even prospered due to his advanced capabilities. His life span markedly increased, with more and more tribes having two generations of adults. This jump in longevity was an important factor in the accumulation and transmission of specialized knowledge to younger generations. Expanding food supplies resulted in larger populations. As populations grew, early man migrated into new areas. HS was making his "great leap forward", the transition to modern man.

Man's new capabilities enabled him to significantly modify his environment. He could now build shelters, eradicate pests, store water, etc. As these changes were made, his culture evolved to support them. However, the changes were much too rapid for biological adaptation driven by natural selection. In effect his physical traits were frozen as he begins to radically modify his food supply, the tools he uses, his social behavior, and the collective knowledge of his community. Miss matches became inevitable. This new capability for advancement presents a new challenge unknown to any other species.

Adam in Africa was a survivor of rapid climate change and the ancestor of all Homo Sapiens. His culture became his survival trait, just as claws and sharp teeth were the survival trait of his animal predators. The development and transmission of cultural information, supported by a rich language and advanced cognitive functions, broke the mold for change. Over Time, the consequences would affect all subsequent life forms.

Cultural Library

Before, we described a library of accumulated genetic knowledge. It contains the blueprints for all existing life forms, 'species knowledge' that has evolved over billions of years. Over the past 5 million years we have added to this library, notably the expansion of the books covering certain upright primates. One, Homo Sapiens, has new material on how to build a large, highly structured brain. Of course, this library does not really exist. Genetic information is real and encoded in the DNA of living organisms…we just don't keep it in libraries. It is of use only to the species and if the species goes extinct, the knowledge is lost.

A distinguishing element of Homo Sapiens is that his species knowledge includes genetic instructions that support cultural transmission. It would be convenient to say he has a culture gene, but things are much more complicated. First his genetic inheritance supports social behavior, some of which is shared with all mammals. Second, he is capable of learning and using a complex language. While some of this is learned and varies from one community to another, the ability to learn a structured language using symbolic representations and a structured syntax are all a part of his genetic inheritance and are universal to all HS communities. No other species has comparable genetic information. Finally, HS consciousness and concept of self, drives him to greater cultural learning.

Therefore, the real news is the development of an entirely new library, the accumulation of memes (cultural elements) that make up the common knowledge of each community of social animals. In our new library we will devote a book to each cultural community with chapters on various cultural elements like language, historical stories, religion beliefs, behavior norms, technology, music, art, etc. Up to recent times, the cultural communities were hunter/gatherer tribes. The fact that each tribe had chapters on language, religion, technology, etc. demonstrates that they are innate to our species and that our species has limited variation from one social unit to another. Of course, the content of the chapters for each tribe is very different reflecting different environments and different experiences.

In the genetic library, our librarian discarded books on extinct species. In our cultural library, our librarian's task is more complex. Occasionally, entire books are discarded when a community ceases to exist. Somewhat frequently, one tribe conquers or absorbs another. Elements of their cultures would merge, sometimes in unexpected ways. A single language and religion are usually maintained. On the other hand, technology is usually cumulative. Ideally the combined chapter on technology would be the best of both. However, best is in the eyes of the conquering tribe.

Our Cultural Librarian has another difficult task. Memes change in time. Just as technology gets updated with new discoveries, languages evolve, religions sometimes reform, and even histories get revised. The knowledge of a particular culture is always changing; and therefore, the cultural knowledge of our specie is continually in flux. Changes are not necessarily advances; reversals also occur. Real advances often occur in waves as man develops and then transmits his new information stimulating further development by others.

As we will see, our cultural libraries will soon expand at an exponential rate. Soon, written languages, printing presses and massive communications will force us into new ways to record and keep our cultural knowledge.

> *One might say that the first principle of life is survival and advancement by the transmission of inherited traits. This innovation gave provided man with the genetic information to produce complex plant and animal species capable of surviving in a huge variety of environmental conditions. It also gave one species the capability for advancement by cultural transmission that must be regarded as the second great innovation of living organisms. It would shortly impact all of earth's living organisms.*

Those Special Primates

Some Homo Sapiens traits distinguish him from all other life forms. He is a primate with an exceptionally large brain. He is a social animal, and his brain has plasticity to support learned behavior. Uniquely, he has an innate ability to learn languages and he uses language to learn from others, and to educate his young. He is conscious to the extent that he recognizes his own interests and has a developing sense of self. Religions continue to be useful to some and needed by all to deal with the uneasy questions raised by their conscious minds.

Over time some men begin to view themselves as not only separate, but special and above all other forms of life.

> *The first pre-man species was a primate who walked upright. Five million years later Adam (notional) and his offspring, had essentially*

all the physical traits of modern man. This change, gorilla to man, was driven by natural selection. Man's enlarged brain was the single most important feature of his evolution. Notably, it developed synergistically with his culture, language, and his newly discovered consciousness. Our Adam and his offspring inherited the capability for a rich language and culture, which became the driving forces for future changes.

CHAPTER 4

——— ✺ ———

Civilized Man

Homo Sapiens (HS) first appeared in Africa about 200,000 years ago. Even at that time he had a rich genetic inheritance based on millions of years of tribal living. His genetic code was relatively complex; it enabled cultural transmission of both information and knowledge. Unlike any other animal species, his culture drove his further development. At times, his consciousness led him to reject life as it was, even if it incurred the wrath of the heavens. Increasingly he became willful, curious, insubordinate, even reckless as he sought novelty.

Out of Africa- Verse Two

Early man (Homo Sapiens) migrated to Europe and Asia, just as Tribal Man had a million years earlier. With his new skills, social and technical, he emerged as the dominant species on the planet. His numbers grew. He adapted as he populated lands with different environmental conditions. He migrated from the African tropics to the moderate climate of Western Europe, the steppes of Europe and Asia, and the tropical heat of India and Indonesia. His food supply and available shelter was different in each area. As he migrated, he learned and adapted.

The last remaining pre-man specie, Neanderthal Man, disappeared from Europe about 12,000 years ago. He survived for almost 200,000 years in Europe during an ice age. The Neanderthals were cave dwellers and excellent hunters. As the climate warmed, they did not adapt well and went into decline. The decline accelerated as early man migrated into his territory. While Neanderthals were strong, they lacked adaptability and agility.

Early man migrated to North America from Asia, crossing the Bering Strait. These men were adaptable and had advanced hunting skills. They domesticated wolves, which became their companions, the first known instance of a domesticated animal. They became known as dogs. Dogs and early man were symbiotic; man gained a keen sense of smell and dogs a reliable source of food. Within 1000 years early man had populated all of the Americas, the fastest blitzkrieg in the history of the species. It contributed to the extinction of many native animal species…some were easy prey.

> *By modern man's standards, early man was primitive. However, by the standards of history, they participated in what was man's first great affluent society. They were excellent hunters and had ample sources of food. They existed essentially on all continents. They lived in tribes with limited external contact. Within the tribe, they were socially well adapted, and they had rich tribal cultures. Significantly,*

they did not practice agriculture as we know it and their populations were relatively stagnant. Their hunter-gatherer lifestyle proved very stable and would continue in isolated tribes into modern times. However, change is coming; most men are about to get civilized.

Throwing the Dice

Early man lived well; the men hunted game and the women gathered locally available cereals, fruits, nuts, and vegetables. This works out to be a good mix of high protein on occasions and, together with, a steady stream of foodstuffs from plants. Most lived on what we now call fertile land, near rivers or lakes, and in temperate climates. Whether Asia, Mid East, Europe, or the Americas they adapted to what was locally available. They lived so well that occasionally they decimated the local pray and exhausted the land. When their food supply dwindled, they moved to another area.

Missing from our contemporary view of early man was his savagery. Far from being noble, they were in almost constant warfare with neighboring tribes. Death rates were high particularly for men. A common female fate was abduction as a sexual prize. Population increases usually led to more intense competition among the tribes. Violence kept population densities stable...tribal populations increased only if there were limited migration opportunities.

About 10,000 years ago, give or take a few thousand years, men in different regions embarked on an experiment called agriculture. Some say that man and the planet have never recovered. Farming brought population explosion, new disease, and many believe, the beginning of environmental deterioration. Agriculture did not improve the lives of most men of the time. While he lived longer, he worked harder at more menial tasks. His tribal structure began to deteriorate often giving way to brutal political structures. And significant inequalities developed; even, for the first time, slavery. In some ways he traded mortality for morbidity.

Collectively did man make a bad decision? It is not likely he even made a choice. Agriculture developed in regions that already had relatively large populations. Many believe that population pressures forced him into more effective ways of obtaining the food he needed. Some would say he was driven to invent agriculture by his incessant innovation. However, if that were true, why would an isolated community retain its hunter/gatherer lifestyles for thousands of years. It is likely that population pressures in fertile areas drove man to agriculture to increase food supply and mitigate the violence of food shortages. Few men had choices, and none could have imagined the consequences of gradual urbanization. More than likely, it was another inevitable step on man's evolutionary journey.

> *Man lived successfully in hunter/gathered tribes for hundreds of thousands of years. Then he threw the dice. Abandoning his tribal ways, he gambled everything on an experiment that we now call civilization and the challenge of urbanization. It was likely not a choice, but an inevitable change driven by his accumulated knowledge. It was the*

> *most significant cultural change in the history of man. However, for thousands of years after the event, most men were less well off than before.*

Farming and Urbanization

The age of modern man really began with an innovation, the domestication of plants and animals. As early as 10,000 years ago, a few communities first produced food by farming. It occurred independently in several different places, including the fertile crescent in Mesopotamia, the Nile valley, and the Indus Valley in India. The availability of suitable plants and animals for domestication was the dominating factor in where farming occurred. Farming supported higher population density and led to larger communities. Once established, farming communities grew in population and then expanded into new areas. When they encountered, hunter-gather tribes, or other farming communities, there was conflict; the larger communities were usually the winners. In some places there were natural (geographical) barriers to migration. This allowed some tribal societies to survive for thousands of years as the rest of the world became urbanized.

Farming brought man into closer contact with animals. He was exposed to new microbes, some of which made him sick. Over time, the exposed populations gradually develop some immunity to the new diseases. As they migrated into new areas, they met tribes without immunity. Disease often decimated these tribes. For thousands of years, tribal devastation was due more to diseases than to hostilities.

The development path which led to what we now call a civilized community was repeated time and time again. First was the domestication of plants and/or animals. The most common domesticated plants were the same cereals we have today, wheat, oats, rye, etc. These crops could be seeded, grown, harvested, and importantly, stored for future use. Domesticated farm animals were generally large, docile herbivores, goats being the first. As domestication proceeded, food supplies would exceed those available from a hunter-gather lifestyle. Increased food supplies results in higher population densities that in turn promoted even more food production. Also, women, freed from tribal hunter-gatherer roles, raised more children.

As communities grew, they became too large to functions with the egalitarian relationships of tribal communities. Competition and stratification within the community occured with most men producing food and a few exercising food distributions. Man became tied to the specific lands that he cultivated and was less inclined to seasonal migration. With less migration, he began to accumulate personal belongings and accumulated goods became an indicator of success. He often traded for goods he did not produce himself. Bartering became a part of many men's lives.

Political structures emerged usually taking the form of autocratic rule by a single person (chief), or a small elite group. The chief was a permanent, centralized authority, and he made all key decisions. While some men became a part of a new warrior class, most remained tied to food production; they paid tributes to support the leadership. Communities grew to number in the thousands and for the first time in man's history, individuals encountered people they did not know without attempting to kill them.

A small class of elites developed; they were positioned between the leader and most tribe members forming a hierarchy both in terms of their importance and their access to goods. They began to enjoy luxuries not available to all. Stratification impacted the lives of all within the communities, many not for the better. The warrior class would not only defend the community's territory, but also enforce the community structure. The egalitarian tribal ethics nurtured by millions of years of tribal living was usually replaced by rigid hierarchical rule.

The early political structures (government) introduced a dilemma fundamental to all centralized communities. At their best, they provide services to the entire community, services unavailable without the political structure and a government. At their worst, they vest essentially all power and wealth in the hands of a few, usually the elite. These functions, both noble and selfish, are inter-linked and at the same time in conflict. They frequently are the basis of internal unrest. In essentially all early urban communities, unrest was brutally suppressed by the hierarchy. Throughout history there have been many attempts at communal societies based on egalitarian principles. What they have in common is they all failed due to the internal tension caused by individualism. Sharing and the division of goods, relative status, sexual favors, all created tensions. Feuds were common and often resulted in the breakup the community. There are clearly limits on the power of culture to change human behavior.

While agriculture initially emerged in just a few

locations, it spread primarily by migration. When horses were domesticated (about 6000 years ago) migration accelerated; for the first time, man could travel faster than his legs could carry him. Shortly thereafter, migration was further enhanced in some communities by a new tool, carts with wheels. With more rapid migration, larger communities absorbed many smaller ones, and assimilated their cultures.

There were now many opportunities for competition among men within a community and a few men, usually the community leaders, were able to make choices based on their self-interests. Man, not only competed for food, land, and goods, he competed for leadership, social position, and sexual favors. There were winners and losers and great inequalities developed. Thus, civilized man's first governments in most urban areas were brutal hierarchies, where many members of the communities were no better than slaves. From a political view there were no human rights and women were seldom a part of the ruling elite.

Agriculture brought increased food supplies and larger populations. Man became tied to specific lands and was less inclined to seasonally migrate. He traded for goods he did not produce himself. With less migration, he began to accumulate personal belongings. Successful communities produced required goods with time to spare. Time was spent advancing agricultural tools and techniques, building shelters, and erecting monuments. This led to the division of labor, societal stratification, elitism, and, without exception,

> *inequalities among men. Larger communities developed autocratic political structures; brutal leadership was the norm This contrasted with prior tribal communities, which were largely egalitarian with a 'fatherly' leader.*

Early Civilizations

As early as 7000 years ago there were urban communities in Mesopotamia; it was believed to be the center of the agriculture/urbanization revolution. The Mesopotamian Empire[13] encompassed the communities in the Fertile Crescent (Iraq) and stretched to the Mediterranean Sea. Like many to follow, it was oppressively ruled by terror and pillage.

About 5000 years ago tribal communities along the Nile River were united to form Egypt, the greatest of the early civilizations. The empire grew in size and importance for about 2000 years and then declined for another 1000. Their agrarian economy was prosperous by early standards resulting in wealth...and surplus labor. Driven by their god-king leaders they were the first to become great builders of temples and royal tombs. Of the great wonders of the early civilizations, only the pyramids survive.

Likewise, Indus-Harappan (Indian), and Chinese civilizations formed and prospered. As the agricultural revolution continued over the next 5000 years, other smaller

---

[13] Empires are states controlled by a single political entity. They are often bound by force and may not share a common culture.

empires emerged and spread over most of the planet. A few tribal communities continued to exist in isolated areas.

Of great significance, some cultures learned to write, an innovation that allowed highly trained scribes to document and communicate their culture in both space and time. The earliest we know of were written accounting for goods and services. Only the last 0.1 percent of man's history has there been any written documentation; the records of the other 99.9 percent are captured in his gene pool, in the structure of his extraordinary complex mind, and in his evolved culture. Finally, our cultural library has written as well as symbolic content. However, early writings, were highly tainted by mystical views. Most early written histories are stories of man's beliefs rather than records of events. The exceptions were the written accounting for goods and services.

Early religions continued to evolve; increasingly, man needed to reconcile questions and uncertainties that come with self-awareness. As communities grew and the leadership became firmly established, another phenomenon occurred. The political leadership claimed to be divinely anointed and even declared they were gods. Along with this, their leadership roles were often inherited by their offspring. Early "civilized" man tended to be ruled by men who inherited their power and influence and thought themselves gods. The practice of inherited power lasted for thousands of years.

From the beginning, urban communities were stratified. Most men worked hard and had few choices on how they lived; their lives were controlled by tradition and their leadership. In many ways they were captives of the urban community if not outright slaves. On the other hand, a small but significant percent of the population was now able to routinely make choices about where and how they lived. They had the mental tools to formulate alternatives, and they acted accordingly. Some were able to exercise great power over the environment and other members of their community.

Often, we view the beginning of civilization as the beginning of mankind's history. Actually. we are marking the beginning of a new lifestyle, farming/urbanization, which displaced the 3-million-year-old hunter-gather lifestyle.

Moral Religions

With urbanization came larger communities where a man no longer knew or even recognized many of its members. In prior times those he did not know…were his enemies. Further, he often did not know his new leaders who increasingly were brutal when dealing with the community… very unlike his tribal chief. And, of course, he continued to have unanswered questions about his origins, his purpose, and his mortality.

Before urbanization, Tribal Man's religious beliefs had rituals focused on death, burial, and/or the afterlife. Morality or even guidance for man's behavior was not prominent nor was it needed. Over millions of years, tribal man evolved a morality (innate and cultural) that was embedded in tribal living. He knew instinctively the behavior that supported tribal harmony. Those outside the tribe were enemies and their welfare was of no concern.

However, beginning about 3000 years ago, a common phenomenon occurred in many of the urbanized civilizations. New religions emerged that differed significantly from their predecessors. The new religions had a moral purpose; prophets, and deities that possessed a moral vision. Some taught ethics for this world and others tied current behavior to hopes and fears about what happens after death.

The moral content of the new religions was that all men should be treated humanely. This suggested that man should extend his circle of acceptance to men outside his immediate family and friends. Extending himself to those he did not know or recognize, was contrary to his instinctive tribal behavior. The fact that the cultures of many of the major civilizations of the time would evolve ethics counter to their biological inheritance seems miraculous. The new morality, of course, was taken differently by different men. It is also fair to say that the acceptance of "others" was seldom all-inclusive. Never-the -less, over history, the challenge of "civilized man" has been to extend his circle of acceptance to larger and larger communities.

The moral leadership for this change was by truly inspired men; and the acceptance of their message says a great deal about the need for changes within their communities. The fact that these great religions (Hinduism, Judaism, Zoroastrianism, and later Confucius, Buddhism, Christianity, and Islam) had common elements suggests they related to early-civilized man's basic needs. Religious memes, like all others, only replicate when accepted and supported by elements within the community. A religious sect forcibly imposed on a community does not have traction until it is internalized by the community.

Judaism, Christianity and Islam have a common core sometimes called the ethics of Abraham. Unique at the time, Abraham committed to a single God. This is a dramatic change from tribal religions. While not unique to Abraham the meme of monotheism became the norm for future cultures. With multiple gods, most any belief or behavior was possible depending on the choice of god. With one God, the choices were narrowed. Some argue this moves the community towards increased accountability (for both man and his God).

Religions, like language, have always been an important cultural element and are often the glue that holds communities together. The ethics of Abraham also supported individual rights appropriate to human dignity, and freedom for the individual to choose their God without coercion from the community. These qualities… monotheism, human liberty, equality, and fraternity are at the heart of these great religions. In Persia, Zoroastrianism was founded on similar principles.

As the new religions gained acceptance, there were individuals and/or groups who maintained and interpreted the practice of others in their religion. Inevitably these men also altered the scope, content, and practices consistent with their own beliefs and in ways that extended their tenure and influence. Often the early religions of different communities had similarities that over time were channeled by their tenders into quite different practices. This led to even more religious diversity.

As with other evolving phenomena such as language and consciousness the enablers of ethical behavior evolved over time. Early on, most men did not have sufficient cognitive ability nor cultural freedom to act morally. Nor did they have a sense of agency. The lives of most men were driven by the will of other forces such as their gods or their rulers. It will be a long time before the average man has true religious freedom.

The religious thinking of Confucius Buddha, Zoroaster, and the Jewish prophets, together with the Greek poets, artists, and philosophers, made the 6th century BC (ca. 2500 years ago) a pinnacle in the development of human wisdom. This century is called the Axial Age, the time in history when a few men, truly the elite few, broke the bonds imposed by man's biological and cultural inheritance. Their awakening was enabled by an unencumbered, conscious, mind.

Finally, a few loudmouth apes had something to say! Inspired men brought a new message to early civilized man, the message of morality. Their teachings were embraced by many and formed the basis of the great religions, most of which still exist today. Their influence on history far exceeded that of even the most successful generals and politicians. However, most men were still oppressively dominated by cultures that reinforced the power of a few and maintained the inequalities that have existed since the beginning of civilization.

The Minoans

Only recently we learned that the first prosperous European culture was that of the Minoans. They inhabited the Island of Crete and several other islands in the Eastern Mediterranean. The Minoans rose to prominence some 4000 years ago and flourished for at least five centuries. It was a civilization of sophisticated art and architecture, with vast trading routes that spread to Egypt and the neighboring Greek islands. They built elegant cities with multi story buildings containing many highly decorated rooms. Their homes had water closets complete with running water and sewers. They were far advanced of the Greeks and, in many ways, the Egyptians.

They were the first civilization to have a navy. Minoan prosperity was based on large-scale trade that ranged from Sicily, Greece, and Asia Minor to Syria and Egypt. The Minoans employed the first ships capable of long voyages over the open sea. Chief exports were olive oil, wine, metalware, and magnificent pottery. This trade was the monopoly of an efficient bureaucratic government under a powerful ruler whose administrative records were written on clay tablets, first in a form of picture writing and later in a syllabic script. As neither script has been deciphered, our knowledge of Minoan civilization is still limited.

The glory of Minoan culture was its art, spontaneous and full of rhythmic motion. Art was an essential part of everyday life and not an adjunct to a religion or to the state. What little is known of Minoan religion also contrasts sharply with religiosity in the Eastern Mediterranean. There were no great temples, powerful priesthoods, nor large cult statues of the gods. The principal deity was Mother Goddess; her importance reflected the relatively high position held by women in Cretan society. She was probably the prototype of such later Greek goddesses as Athena, Demeter, and Aphrodite.

About 1500 BC, the date is uncertain, a huge volcanic eruption heavily impacted the Egyptian, Greek, and Minoan populations. The blast and tsunami that followed essentially destroyed the Minoan civilization. Survivors remained for several generations, but the culture was decimated and never recovered. Thus, a volcanic blast enabled the rise of the ancient Grecian culture and our Greco-Roman heritage. Man's destiny continues to be influenced by the uncontrollable.

> *The Minoans were the first known European/ Asian civilization. They were advanced relative to early Grecian civilizations . They prospered using their "Navy" to trade goods in the Eastern Mediterranean. Their 500 years of prosperity was devastated by a volcano in 1500 BC.*

Europe's Golden Past

It was the epoch-making discoveries of the English archaeologist Sir Arthur Evans that first brought to light an **early Grecian civilization**. Their existence had previously been hinted only in the epics of Homer and in Greek legends. Evans unearthed the ruins of a great palace at Knossos, the dominant city on Crete after 1700 B.C. Rising at least three stories high and sprawling over nearly six acres, this "Palace of Minos," built of brick and limestone and employing unusual downward-tapering columns of wood, was a maze of royal apartments, storerooms, corridors, open courtyards, and broad stairways. Furnished with running water, the palace had a sanitation system that surpassed anything constructed in Europe before the Roman times. Walls were painted with elaborate frescoes in which the inhabitants appear as a happy, peaceful people, with a pronounced liking for dancing, festivals, and athletic contests. Women are shown enjoying a freedom and dignity unknown elsewhere in the ancient Near East or classical Greece. They are not secluded in the home but are seen sitting with men and taking an equal part in public festivities, including bull fighting.

The Greeks settled the Asiatic coast and offshore islands prior to 500 BC. They developed a sophisticated culture placing a high value on intellectual pursuit and the arts. More than any predecessor, they began to think of space, time, and the state of man. They essentially invented philosophy. As they accumulated wealth, some Greeks had leisure time, and the freedom to pursue the good life. At least the elite had that opportunity; most Greeks of this period had neither wealth nor freedom.

The **Persian Empire** expanded into Turkey in the North, Afghanistan and India on the east, and across North Africa in the West. It was the greatest empire the world had seen to that date. Zoroastrianism, a monotheistic religion, was the defining element of their culture. The Persian rulers[14] were one of the first to show any humanity toward their subjects and toward those they conquered. As they expanded

---

[14] Cyrus the Great and Darius I

westward, they collided with the Greek-speaking world. This brought, for the first time, a moral dimension into the world of politics…a contest of ideologies. Most future wars would pit men and their gods against one another.

"The glory that was Greece," in the words of Edgar Allan Poe, was short-lived and confined to a small geographic area. However, it influenced the growth of Western civilization far out of proportion to its size and duration. The Greece that Poe praised was primarily Athens, during its golden age in the 5th century BC. Strictly speaking, the state was Attica; Athens was its heart and soul. The English poet John Milton called Athens "the eye of Greece, mother of arts and eloquence." Athens was the city-state in which the arts, philosophy, and democracy flourished. It was the city that attracted those who wanted to work, speak, and think in an environment of freedom. In the rarefied atmosphere of Athens ideas were born about human nature and a political society that are fundamental to the Western world today.

The Greek-Persian wars were long and brutal. Interspersed were wars among Greek states, perhaps in part inspired by Persian bribes. These internal wars revealed an unflattering truth about the Greek city-state of Athens. Athens created an empire and, in the process, turned her allies into subjects. While proud of her own freedom and independence, she snuffed out the freedom of others. The meme of democracy was short lived; it was too fragile for the civilizations of the time. By 390 BC both Athens and neighboring Sparta went down to defeat.

A Macedonian King, the father of Alexander the Great, led to a comeback of the Greeks over the Persians. He first united Greece; and 2300 years ago, his son became the first warrior-political leader to open the option of a world empire. He first consolidated his rule in Eastern Europe, conquered the Persian Empire and extended his rule further east into Asia. His leadership and conquests have not been matched by anyone since. He established a great learning center in Alexandria (Egypt) where works of all cultures were assembled with the finest scholars of the time. Years later, there was a significant cultural extinction event when the library and its contents burned to the ground.

Hierarchical organizations were usually held together by brutal responses from their leader. This was viewed necessary to overcome innate tribal behavior. Greece was the origin of two very different behavioral memes that have survived: Sparta Teams and Athenian Pluralism.

> Sparta devised an alternative organization that nurtured cooperative effort that overcame biological and tribal behavior. They formed communities or teams of young men who were committed to their group rather than their tribe or family. The Sparta model was very brutal; the entire Sparta nations was organized to breed, train, and deploy highly effective warriors with no regard for family bonds. The model of individuals from different communities, giving up their individuality and working together for a common group goal is the surviving meme. Modern day militaries use a highly disciplined version of the Sparta model; almost all effective organization utilize a team effort to accomplish well defined goals.

While Sparta is characterized as brutal uniformity, Athens was the hatchery of diversity. Athens encouraged trade outside their city/state and interactions with others, which inevitably spawned subcultures within the Athenian community. Unlike Sparta, this diversity could exist as a subculture within the hierarchy and compete with other subcultures for recognition. Athenians demonstrated pluralism, a concept largely unseen in prior communities.

These two very different organizational memes tested key concepts: authoritarianism versus libertarianism, internationalism versus isolation, and totalitarianism versus democracy. Most civilizations today utilize elements of both.

The Romans in Europe and the Han dynasty in Asia followed. Their troops were highly organized, spreading their influence to surrounding nations, resulting in great empires ruling most of Europe and Asia. The Roman Empire developed two cultural innovations, a democracy for the elite, somewhat modeled after Athens, and a new form of warfare.

Successful Romans were relatively free to pursue their own interests. They selected an emperor who was then given unrestricted war powers and the power to rule the empire. The elite retained personal freedoms. This model would be repeated in European Monarchies with feudal privileges for the aristocracy.

The Romans also innovated a new form of warfare. Instead of destroying their enemy, they dominated the land they captured and introduced the Roman culture…and then converted their former enemies to members of the Roman Empire. Their warriors were added to the Roman Legions. This successful meme survived into the 20th century in the form of European Imperialism.

Other great civilizations emerged and at different times and advanced in different ways. They were built and defended by warriors supported by large urban/agriculture communities. Many fell either to other warring groups or by exhausting their base of support. Those in China were re-built time and again for thousands of years. The Romans were particularly successful with advanced technical and political systems. For almost 500 years they dominated the western world. Unlike the Greeks, they contributed little to man's religious or philosophical thinking. Their subjects did. Christianity came out of Judaism and appealed to Greeks, Romans, and Semites. Over time the teachings of Jesus were shaped and adopted by the Romans; these teachings would survive the fall of Rome in the dark ages and re-emerge as a major force in Western Civilization.

In some areas, communities within an empire fell in the face of repeated attacks by barbaric tribes. These tribes were usually large with ferocious leadership. Their pattern was to attack, pillage and move on, leaving the community without resources and/or leadership. The result was isolated feudal states. Without the warriors supplied by the declining Greek and Roman empires, feudal system evolved, where mini states built and lived in a fortresses for protection. They accepted occasional losses and were sometimes destroyed. Most sustained elements of their cultures including their religions.

While all the early European civilizations declined, much of their art, philosophy, religions, and political structures that emerged from their cultures did not. Like the Axial Age, man was developing memes of enormous importance to all future civilizations. However, our cultural library of memes is just getting off the ground.

Abraham's Other Son

In the 7th century the Prophet Mohammed established a new religion, Islam. It was based on the Ethics of Abraham, a set of values shared by Muslims, Christians, and Jews. For more than a thousand years, Islam provided its subjects, regardless of their heritage, a set of rules and principles for the regulation of their cultural and their personal affairs. Even today, nations with large Islamic communities sometimes have a bond, beyond national boundaries, with other Islamic states. For example, there are Councils of Muslim states, and Muslim blocks in the UN. This is unusual. Muslims often view history as beginning with Muhammad. Even Egyptians, with their rich cultural history, identify more strongly with their Muslim cultures, even celebrating the date when they were conquered by Muslim invaders.

From the 7th to the 12th century, Islam became the dominant culture in the Middle East. It spread throughout southern Asia, the mid-east, northern Africa and parts of Europe. Early-on their culture was pluralistic, characterized by openness and an obsession for knowledge. Scholars in Baghdad set about translating significant scientific and philosophical works of foreign cultures. Over a period of 500 years they led further advances mathematics, astronomy, and physics.

The Crusades, beginning in the 10th century, were not decisive in terms of Muslim Imperialism; however, Muslim dominance of the middle east continued for centuries with Turkish and Persian Dynasties. The rapid and near total success of the Muslim armies convinced them they were truly an instrument of God. Their fortunes began to change when the Ottoman empire was defeated by Christian armies in 1680. While Muslim armies remained, there winning days were over. Perhaps for the best, they were no match for the rapidly emerging Western Civilization.

Islamic religious culture is based on the example of Mohammed. He is praised as a man of peace, but he was also a military leader who announced the concept of the Jihad, a dedicated struggle to achieve the Islamic concept of God's kingdom on earth. Some tenders of the Islamic faith have declared two worlds, Islamic, and non-Islamic. These two worlds are at times in conflict.

Over the first 10,000 years or so of civilized man, all the great empires failed; never-the-less man had made a giant leap. First, of course, he survived; he had defeated and subjugated all other species. He developed rich, comprehensive languages and a high level of self-awareness. He learned how to work together, first within his tribe, and then in large communities. He invented agriculture and put the rest of nature to work for him. His habitat emerged from caves and became great cities. With his new identity, he developed religions based on morality. A few men had well-developed reasoning, and many had choices of goods and lifestyles. He also developed art, philosophy and the be the beginnings of science. Islam was established in the 7th century and prospered for over 500 years.

CHAPTER 5

—◇—

Modern Man

During the dark ages, Islam and China were the most advanced civilizations. Barbarians overran many communities forcing survivors to trade their freedoms for the security of feudal kingdoms. Renaissance and enlightenment followed. Both good and bad ideas emerged, and many did not survive. In the 20th century the British led the industrial revolution and spread its institutions around the globe. USA prospered greatly and led the world through the rest of the century and into the 21st. The world achieved relative high levels of prosperity. Two difficult issues remain; there are major inequalities among men and collectively, men are exhausting their raw materials and savaging their environment.

Europe Awakens

Europe was briefly united under what is known as the Carolingian Dynasty, which lasted for about 300 years. At its peak, King Charlamagne ruled territory that included France, Germany, Denmark, most of Italy and Eastern Europe. These were feudal times when Knights and Lords, ruled the peasants and protected their individual "kingdoms", all with a permissive Catholic Church. The Dynasty ended when Viking marauders first invaded England and then the European mainland.

The Catholic Church inspired the first crusade (1090 AD) to displace the Muslim rule of the Holy Land. The Crusades lasted for over 200 years, resulting in many deaths but limited geo-political changes. There was an unintended consequence of the Crusades, contact between Europeans and Muslim intellectuals. Muslims gave back to the Europeans their rich classical Mediterranean heritage, which was largely lost during the dark ages[15]. We see the best and worst of cultural transmission: Warriors killing for the sake of cultural (religious) dominance in Jerusalem and Christian and Muslim intellectuals exchanging cultural memes in Southern Spain.

The Black Plague began in Central Asia and was carried west by the marauding armies of Genghis Khan. It reached Europe by the 1340s. It was one of the most devastating pandemics in human history. The total number of deaths worldwide is estimated at 75 million people, 20 million deaths in Europe alone. In some cities, two-thirds of the population died. The same disease is thought to have returned to Europe every generation with varying degrees of severity until the 1700s. The Black Death had a drastic effect on Europe's population and irrevocably changed the social structure. It was a serious blow to the Catholic Church and resulted in widespread blame and persecution of minorities

---

[15] The Catholic church through its monasteries largely maintained European culture during the dark ages. They paid scant attention to scientific or philosophical ideas that did not support Christian ideology.

such as Jews, Muslims, and foreigners. The carnage and uncertainty of daily survival caused men to question and often mock and reject their existing institutions. The stage was set for a renaissance and reformation.

The Renaissance[16] spanned roughly the 14th through the 16th centuries. The period is best known for its achievements in literature, art, and architecture. It was driven by a few men with extraordinary talent; they were enabled by sponsors of extraordinary power and wealth. It originated in Italy with Leonardo De Vinci and spread throughout most of Europe. However, it remained a period of economic stagnation with little progress in science. At this point men were showing great imagination, but most lacked the power of rational thinking. And, as with other periods of achievement, most men were unaffected. Kings and Emperors held political control over much of Europe, and the church maintained social order, with proscribed moral behavior…just as tribal culture did for early man. This was done via a vast network of churches, monasteries, and priests aligned with the Roman Pope. A Dutch Philosopher Spinoza noted centuries later, when religion and state are combined, rights are trampled, free thought discouraged, and innovation snuffed out. Only the elite are exempt.

Slowly Europe threw off the yoke of feudalism. Advances in agriculture occurred and a food surplus slowly developed, followed by larger scale commerce. Trade became possible as communities became more secure and affluent. Over several centuries, Europeans propagated their civilization

---

[16] The Renaissance refers to a period of great changes in Europe. The word renaissance is now often used to describe cultural events of historic importance.

(including Christianity), together with commerce and military power. Advances in military technology, including gunpowder, enhanced their efforts.

Martin Luther was a German priest. He challenged the authority of the papacy by holding that the Bible is the sole source of religious authority. He argued that an individual's faith, not the church, secured a man's salvation. His writings were widely distributed in part due to the recently invented printing press. In effect he declared religious independence from Rome, asserting that man did not need a mediator between him and God. What followed was more than religious reformation; for the first time, according to Luther, the common man was not subject to the authority of others on moral matters.

With printing presses, literature became more available; knowledge was no longer reserved for aristocrats and the clergy. And it could not, be controlled by the state. It was now available and used by all those with intelligence and curiosity, regardless of birth, class, or caste. These ideas helped to inspire the Protestant reformation…and changed the course of Western civilization.

While Muslims remained the dominant civilization, Europeans emerge from the Dark Ages with agriculture advances and more secure food supplies. Populations greatly expanded. Centers of commerce developed, and trade flourished. Nations formed around European communities with similar cultures. Wars among the nations were common, usually with insignificant results. The Black

Plague devastated many nations. Kings and other nobility controlled the activity of most men while the church controlled their thinking. A few innovative men made historic cultural contributions in literature and art. Significantly, some men now felt themselves free from the domination of the church, setting the stage for reform.

Enlightenment and Rational Thinking

A period of enlightenment followed the reformation as some men broke free from cultural/religious restraints. New thinking was based on reason rather than superstition and religious edicts. There were huge advances in science and philosophy…both were previously the domain of the church. Art and literature also advanced and ideas leading to political freedom emerged. Not surprisingly, this brought turmoil to many medieval societies. New thinking challenged man's relationship to nature, God, and to the state. New ideas emerged on economics, philosophy, politics, and human rights. While the elite innovated most new cultural memes, all men gained entitlement to think and act on their own thoughts. Not surprisingly, intellectual elites were the primary beneficiaries.

Enlightenment and rational thought brought scientific breakthroughs. Sir Francis Bacon outlined a new learning process through science. Newton de-mystified deeply rooted concepts. For example, the laws of gravity replaced God as the force that control the movement of the planets and the stars. Of significance his work provided an example of how rational thinking can change how men view their world. A steady stream of new science and technology was established that continues today.

In the late 17th century, the English philosopher John Lock argued that a government could only be legitimate if it received the consent of the governed through a social contract that protects the natural rights of life, liberty and property. If such consent was not given, argued Locke, citizens had a right to rebel. His writings influenced Thomas Jefferson in writing the American Declaration of Independence. A new meme for democratic rule was now on the table, this time to stay.

The "Scottish Enlightenment" was a period of intellectual ferment in Scotland running from approximately 1730-1800. David Hume was the most influential thinker; his philosophical investigations inspired Adam Smith and the first work of modern economics, "The Wealth of Nations". Concepts of free trade and free markets, previously achieved by bartering among individuals, gained credibility. Scottish Enlightenment provided the heart and soul of the British Industrial revolution.

Enlightenment also had a profound impact on the fortunes of Western Civilization. The new thinking, technologies, and commerce brought great advances. The Islamic world did not participate in these advances and lost ground.

> *During the Age of Enlightenment and Reason, intellectual leaders of the time regarded themselves as the courageous elite who would lead the world out of the Dark Ages. The*

movement created the intellectual framework for great advances in many fields including philosophy, science, religion, capitalism, atheism, humanism, communism...all had their roots in this period. Further, rational thought replaced superstition, and the scientific method and experimentation was used to understand our natural world. While Enlightenment brought new thinking, the lives of most men were still unchanged... with some exceptions. Enlightenment had not brought change to the political institutions that still ruled the world.

Dangerous Ideas

Memes of great significance evolved during this period of enlightenment, Darwin's evolution, atheism, the blank slate, and communism They were dangerous either because they threatened man's fundamental beliefs, or because they were dangerously wrong.

Evolution

In the 1860 Charles Darwin published "On the Origin of the Species". His theories shocked the intellectual world, claiming that species evolve by natural selection, and man is subject to the same evolutionary forces as all other life forms. Darwin's theories turned nature upside down. Instead of a top-down design, life evolves from the very simplest to the complex. The 'invisible hand' that led to the evolution of new organisms was not a master mind, but the simple idea of 'survival of the fittest'.

Atheism

Prior to 19th Century, essentially all men believed in some sort of god. Atheism was felt to be abhorrent by most Enlightenment philosophers. In the early 19th century, the German Ludwig Feuerbach spelled out the case that man created his God to deal with his own uncertainties and therefore man's religions tell us about our own inner needs.

By the end of the 19th century, atheism was common among the elite in Europe. The general population retained Christianity; however, many rejected the super-natural aspects attributed to Christ. A movement "God is Dead" began in Europe with Nietzsche and many identified with Marxism (atheism, authoritative statism). Conventional religions began to modernize and humanize their belief structure to be more in keeping with natural scientist and the views of the intellectuals. In most cases, the efforts to modernize the religious institutions met the needs of many if not most; others failed and their followers simply abandoned them.

Blank Slate

The blank slate concept, sometimes called *tabula rasa*, is that an individual is born with no innate or built-in mental content, (in a word, "blank") and that his entire intellect is built up gradually from his experiences. It traces from

Aristotle, St Thomas Aquinas, John Lock[17], and other Muslim philosophers The concept flourished in the 20th century social sciences. Fortunately, it displaced Eugenics, which came to be seen as a crime. Generally, scientists now recognize that man's mind is indeed preprogrammed and organized to process sensory input, motor control, and emotions.

The Blank Slate was an attractive vision. If all of man's shortcomings, for example racism, sexism, and class prejudice can be attributed to culture, they could be eliminated by managing the culture. Unfortunately, the concept is fundamentally wrong. Man's character and behavior are profoundly influenced by his genetic inheritance and his very early experiences. Socialists proposed innumerable programs to correctly program man's culture to achieve their desired (Utopian) outcomes. On too many occasions, governments forcibly imposed these programs. At its best, men lost freedoms; at its worst, men died when they were unable or unwilling to conform. Regardless of its morality, the concept is just wrong.

Communism

Karl Marx believed that every aspect of human life and thought should be governed by social and economic factors. (God is simply a projection of human concerns, a result of unjust social conditions.) Religion would disappear when unjust social and economic alienation were eliminated in

---

[17] Lock further wrote that man was free to define his content of his character, however, his basic identity cannot be changed. From this Lock defined natural rights and the right to political freedom.

favor of equality. Marx believed that equality, and therefore freedom could not reach that point without the state controlling political and economic activities (authoritative statism).

Communism is a political and economic system based on some very dangerous ideas, tabula rasa, atheism, and materialism. The goal is to achieve universal equality. Men are not equal and therefore to achieve equality of results, men must be treated very unequally. The economic system replaces the marketplace with authoritarian control over economic decisions. The political system gives a new elite autocratic power. It was tested in prior years with disastrous results. Communism brought death to hundreds of millions of men at the hand of the state. Statism (surrendering individual freedom for the good of the state) has proven to be a sure path to tyranny. It gave us the horrors of both Nazism and Communism.

This is not the formula for success; its broad acceptance, particularly by the intellectual elite is startling.

The Industrial Revolution

By the end of the 16the century, the West had emerged from her dark ages and embarked on a process of reformation: new thinking, new science and technology, and new concepts for commerce and trade. By the end of the 18th century, Europe dominated the world outside of Southeast Asia, and it seemed impossible for others to catch up. The powerful streams of thought, agriculture advances, and new technologies...led to the Scientific/Industrial Revolution. It would impact the entire planet and continue into the

3rd millennium. Europeans led this revolution; however, its roots were in earlier civilizations, particularly the Classical Greek, Roman, and medieval Islam.

The leaders of the new revolution accumulated capital which, together with technology, led to further industrialization. The worldwide expansion of literate Europeans impacted all lands and all peoples. It pushed science and technology around the globe and set out values, ideals, and economic systems that enhanced man's wealth, comfort, and control over his planet. However, not all men benefited. These Western innovations tend to concentrate and control wealth in the hands of a few. This characteristic is not a fault of the Europeans or of civilization itself. In nature, the most competitive survive and prosper and self-interest became the force for change…and the source of material wealth.

The globalization of commerce provided leveraged for the nations with industrial capabilities; none were as successful as England. Her government policy was to build the greatest Navy, stimulate the new technologies promoting industrialization, and vigorously pursue global commerce and influence. This brought her into conflict with European competitors and with nations around the world with whom she competed and wished to dominate. Warfare was endemic. However, wars were limited in scope and fought by agreed to rules. Most wars were not decisive and relatively non-destructive.

The Rise of Democracies

Democratic forms of government were tried by the Greeks, Romans, and others. All but a few failed; history shows that democracies are unstable. Probably the most common reason is that, over time, elite groups from within, tend to subvert the rights of others. Inevitably, governments grow in size and the majority find ways to syphon off the would-be bounty of the minorities. The actual failure mechanism is often internal decent leading to chaos, and sometimes tyranny.

Western Europe had a history of independent thought, a significant cultural meme needed for a democracy to survive. However, most men were still the subjects of imperial states…the lofty ideas of John Lock were just ideas. France put the democratic meme to the test. The French Revolution successfully dethroned the ruling class. However, they were unable to follow through with an alternative political structure and the experiment ended in failure.

About the same time, a second experiment turned out differently…one of Britain's wars produced an unexpected outcome. The founders of the rebellious 13 colonies in America not only objected to British rule, but they also objected to being ruled by the principles that had characterized governments up to that time. They wanted a fresh start; free from inherited rulers (kings and priests). Further, they based their society on the "natural" rights of the individual, junking the prevalent view that the individual existed for the benefit of the community (state, or ruler). Personal freedom seldom occurred in prior civilizations (or in man's millions of years of tribal history.

The three enduring principles that succeeded were 1) individuals have rights because they are individuals, 2) the rule of Law governs the interactions among men, not the desire of individuals or the tyranny of the majority, and 3) governments exist to serve the citizens, not the other way around. While the concepts were not new, they produced dramatic results when established as the founding principles of a new nation. Further, they were isolated from European conflicts, and in a land with immense raw material wealth. The founders, suspicious of government power, built in numerous checks and balances to counter autocratic tendencies.

Within a few decades people flocked to her shores driven by ambition to succeed and/or to find safety from oppression. The United States grew rapidly in population, prestige, and power. Importantly, with both prosperity and freedom, her citizens enjoyed choices unheard of in the rest of the world. While idealistic and based on values rather than force, she displayed acute practicality. She did not propagate her beliefs to foreign nations and thereby avoided the wrath of superior powers. And she preserved her national unity by allowing regional differences to exist, even to the point of allowing slavery longer than any other major nation. When slavery could no longer be tolerated, she endured a Civil War to preserve the Union without slavery.

For most of the 19th century, the US was the only nation of prominence that had political institutions supporting freedom of thought and action. Then as now, individual freedoms were limited to choices available within their culture. Most men were Protestant, but before long both Catholicism and Judaism were accepted in their pluralistic society. Capitalism drove economic advances; some gained great wealth, but great inequalities remained. Generally, there is prosperity in cultures where individuals are free.

As should have been expected, advances in science and technology again dramatically increase the food supply. Medical advances saved many lives, removing disease as the major check on population growth. For cultural reasons, neither Marxism or atheism arose in the same way in the US as they did in Europe and elsewhere. The lack of a class system and a prosperous free economy defanged arguments for socialism.

In time man began to sense his negative impact on his environment. However, his advancing science was a narcotic that led him to believe he could solve all problems, given time and resources.

Technology, weapons, and natural wealth are important factors leading to success in the newly industrial world. A new factor emerged of even more *importance, individual* freedom was *supported by the culture and political institutions. In prior cultures a few men were free; through them we had glimpses of what man might become. Now in a few cultures, all men were entitled to freedom; they prospered beyond belief. Individual freedom, the newest product of man's conscious mind, emerged as a gift for the 20ᵗʰ century.*

A Century of War

In the mid 1800's the British became the most powerful nation on earth. Her power was based on industrialization and her navy. Napoleon was a military genius who had a powerful army that at times ruled the continent. While he conquered most of Europe, he exhausted France of men and materials. British products continued to travel across every sea, supported by her navy and financed by British banks.

Early on the British knew why they led the world in technology and commerce. However, after many years, success brought complacency. The Germans and Americans began to rival her technical lead. The British experienced a great depression in 1870's; they had saturated the world with their line of goods and services and competing nations began to make products cheaper. Never-the-less the British maintained the illusion of prosperity. The British Empire was at its peak. There was relative peace in the world and many intellectuals believed that war was no longer a viable tool for change. In the early 1900's political leaders began to talk about world peace and prosperity. The US had become a successful nation, if not a world power, with wide recognition for her political ideas and her growing industry. Of the many nations of the world, only about 10 could be considered democracies; half of them were former British colonies.

While the British recovered from the depression their domination of commerce slipped with the Germans being the primary beneficiary. The downward trend continued into to World War I and beyond.

The outbreak of war in 1914 was a surprise...however it

quickly consumed the men and resources of many nations. The fighting was more devastating than in any prior war. In effect, it lasted until the end of the century. Fighting ceased in 1918 with what turned out to be an uneasy intermission. The economic collapse that followed led to totalitarian rule in many of the distressed nations (both winners and losers). The war resumed in 1938 and continued to 1945; by this time US, isolated from overall destruction of WW II, had become a world power with demonstrated manufacturing capability and a respected military. As did the Soviet Union.

What followed were disagreements over the control of defeated nations; this resulted in a protracted "Cold War". However, the differences were much more than territorial; for the first time US was projecting a political view that one nation should not dominate another. This view ignored many inequalities that existed around the world and the economic differences between US and Russia.

The Soviet Union championed a political structure that claimed to represent the common man in his rebellion against economic domination. They exalted man's labor when, at this time in history, man's mind was the key to progress. Statism ruled and individual freedoms were discounted. Their economic system was deeply flawed, and they could not keep pace with economic systems rooted in freedom. In time their leadership became tyrannical and was overthrown.

All of the great powers were belligerents in the wars that raged throughout the century; most were ruined by it…all except the US. While Britain never lost a war, her empire was dismantled. For hundreds of years, European nations had set out, first to explore the world and then dominate it with their military and commerce. This expansion was fueled by the scientific/industrial revolution. At the beginning of the war, the US was a debtor nation with a small navy and almost no army. When it ended, the US was the wealthiest nation in history, the center of economic power, and the only remaining military superpower. At the beginning US wanted isolation from European conflicts. In the end her armies were deployed (usually welcomed) over the entire world.

And yet US kept no occupied lands for herself and aided her defeated allies and former opponents in re-building their societies. If anything, the war spurred the scientific/industrial revolution with the US now on the leading edge. Advancements were dramatic. Electricity and gasoline, both based primarily on fossil fuels became the source of industrial power, later to be supplemented by nuclear energy. For the first time, communications traveled faster than man with the telegraph, followed by radio, TV, and the Internet. Computers were invented, miniaturized, and then embedded in every device imaginable. Weapons of mass destruction were developed and used. Mutually assured destruction deterred war among the major world powers. Again, war became unthinkable to most world leaders.

Food supplies expand with the green revolution (chemical farming). Bio-engineered foods promise to further enhance food production by "naturally" protecting food plants from predators. The world population again exploded. Plants and animal species were disappearing at an alarming rate, most by the destruction of their habitat. It would seem, that given free choice, man is destined to destroy his habitat.

Free markets and capitalism proved to be the winning

economic system. The reason became clear: individual freedom and free markets are winning memes. Feudalism, Mercantilism, and various forms of socialism all failed over time. However, there are some systemic problems with free markets, such as widespread inequality and environmental damage. Inequality is a natural outcome of capitalism... however further excesses occur when individuals or groups derive favorable, government imposed, restrictions in the marketplace. Another problem is that it is difficult, if not impossible to adequately cost environmental damage. A better approach is to legislate limits on the allowable damage. Nations deal with these problems in different ways and one can only say the problems are still with us.

The great religions have been long dominated by the religious elite, the tenders who are trained to control the content and practices of their followers. They continue to satiate the masses; however, cracks appeared as man's science challenged more and more of their religions content. Increasing awareness of science and the evolution of life, resulted in less acceptance of religious stories that served prior generations. Some believe technology to be the new god; they no longer need the crutch of religion. An increasing number of men view religious history as stories that were interesting but that have no divine meaning.

The 20th century, often called the American Century, produced great technological changes which spread around the globe. Man's thinking also evolved, giving higher value to economic and personal freedoms. Great wealth was generated and accumulated by a small percent

> *of the world population leaving immense inequalities. Nevertheless, man created a truly affluent society relative to any that existed in prior times. Yet man remains troubled. His instincts are still tribal, inappropriate for the world he rules. The demise of many of his moral religions leaves many without a moral rudder.*

Man's population now exceeds 10 billion. With modern communications huge inequities are now known to all. Many natural resources are depleting, and our environment is degrading. Biodiversity is narrowing due to the destruction of the habitat of many species. And technology is still growing rapidly, raising more social issues. Some cultures are unwinding as various religions are no longer honored by large segments of the populations.

For the first time an animal species, man, is capable of being an extinction event, an event that would certainly destroy him, as well as threaten all other life forms. It might take the form of nuclear weapons or biological terrorism, or a genetic engineering blunder. To say nothing of potential "natural" causes such as a super pandemic, collision with an asteroid, and/or another major ice age. If it were to happens now, mammals will have ruled for less than 0.2 percent of the history of life on earth…man will have been a very short-lived species. He was full of sound and fury but just another dead end on the tree of life.

Libraries of Knowledge

Our library of genetic information, encoded in the DNA of each specie, has not changed much in the last 100,000 years. What exists was sufficient to spawn incredible changes that have impacted essentially all living organisms. Before, we found little that was special about early man. Now it is undeniable that there is something very special about 20th century man. He is changing this world at a blistering pace with little regard for the destiny of life including his own. Is there something wrong with man's basic character that results in suicidal behavior? Or, given a fresh start, would man make better choices?

Looking through our cultural library we can identify cultural winners and losers. Our measure of significance is the memes impact on the content and quality of where we are and what we believe today:

Agriculture was a key cultural learning that initiated the urbanization/civilization period. It continues to advance and has allowed population expansion beyond natural limits. The transition from tribal to urban communities and farming was the event that drove most of our population growth which in turn lead to leisure time and to cultural memes.

Religion has been a fundamental part of every culture; most were helpful. However, a few have been genuinely harmful to their followers. Monotheism proved particularly important and may have contributed to memes favoring the equality among men. When moral religions emerged, they pushed man to enlarge his circle of acceptance to include those outside his own group. On the other hand, atrocities and even wars have been waged in the name of Religions, often driven by perverted views of the original religious teachings.

The use of the scientific method and rational thinking led the Industrial Revolution. For it to occur, man had to eliminate cultural barriers to free thinking. On the negative side, some men are loud and speak freely, but have nothing to say.

The development of economic, political, and individual freedoms gave man choices that he used to produce great advances, inequalities, and, at times chaos.

Where Are We Now?

So far in the 3rd millennium is more of the same. With major biological and cultural advances continuing, our library of knowledge is in constant turmoil. In the twentieth century, 2 of the big 3 issues were settled. First, democracies with pluralism, are preferred over dictatorial rule. Second, free markets win out over statism and free trade benefits all. The third question is how should men live; what are his values and purpose; and will his God survive secular humanity? The jury is out.

Globalization is now here, and it is a stunning affront to our tribal core. Even after 2500 years of moral religions, we have not learned to live well in large groups. Secularism does not help with its drive to cast out formal religions, it may well make things worse.

Many contemporary religions have fractured into two distinct camps. Fundamentalists hang on to their original religious mysticism resulting in a very limited outreach to others. Increasingly, they rely solely on faith to counter new revelations from science. Religions that have modernized accept modern science; and they worry about the human condition. However, they lack the personal involvement offered by the fundamentalists and many of their elite have converted to atheism. That said, in the US, the fastest growing religions are the fundamentalists: charismatic Christian faiths and un-reformed Moslems.

The success of capitalism and free markets has resulted, as it must, in inequalities. Man's preference for equality of outcomes is innate to his tribal nature. Thus, there is a conflict within man between capitalism and his egalitarian instincts with no obvious solutions. We must recognize that responsibility comes in layers. Is it enough to act responsibly just to ourselves and our neighbors? How about our community and the rest of the world? And other living organisms now and in the future? Can the negative aspects of capitalism continue to be ignored?

Odds say nature may solve this problem with another extinction even. In the past starvation and disease have limited the population of specific species. That could also happen to us. Or the problem might be solved with a colossal biological or nuclear mistake. Of course, there is a better way; we could freely limit our population to levels that our planet can support. Unfortunately, that would take a new awareness and level of responsibility that does not currently exist. And it must be shared by all the world's people.

In summary, after thousands of years we are getting an idea of our own identity. For a few hundred years we have known we are mammals and for a few decades we've understood in considerable detail how we evolved. We are outnumbered on this planet by the ants and outweighed by bacteria. However, our capability for generating knowledge gives us powers that dwarf the powers of all the rest of life on the planet combined. Now for the first time in its 4 billons year history, our planet is protected by far-seeing sentinels, able to anticipate danger from the future. Have we (collectively) matured enough to take on that responsibility? Like it or not we have become both the planets nervous system and its guardian. We need to act that way. If we do not man may well be the cause of the next extinction event…that could wipe out our cultural library and totally decimate our genetic knowledge. There must be a better way.

Man has always been hostile to groups other than his own. Fortunately, with advancing cultures the size of the groups that can live together in peace has jumped from a hundred or less to hundreds of millions. However, man still feels those outside his group are not "real people". There is hope that someday we will be more tolerant of other groups.

Man's progress is demonstrated by comparing how the victor of a competition treats the defeated:

- Early man (hunter-gatherer)- Kill everyone outside our tribe
- Early civilization (Agrarian)- Enslaves outsiders

- Early Urban Empires (Greeks/Romans)- Assimilate them
- Early World Powers (British)- Politicize them
- Current World Power (USA)- Re-build them

Life has developed along a long torturous path. Our motto should be 'all life has value'.

End